Contrastive Analysis of English and Polish Surveying Terminology

Contrastive Analysis of English and Polish Surveying Terminology

By

Ewelina Kwiatek

CAMBRIDGE
SCHOLARS
PUBLISHING

Contrastive Analysis of English and Polish Surveying Terminology,
by Ewelina Kwiatek

This book first published 2013

Cambridge Scholars Publishing

12 Back Chapman Street, Newcastle upon Tyne, NE6 2XX, UK

British Library Cataloguing in Publication Data
A catalogue record for this book is available from the British Library

ISBN (10): 1-4438-4410-1, ISBN (13): 978-1-4438-4410-9

TABLE OF CONTENTS

List of Illustrations

LIST OF TABLES

PREFACE

This book is based upon my PhD thesis at Swansea University, submitted in February 2012. Initially, I would not have thought of undertaking my PhD had my supervisor, Pius ten Hacken, not encouraged me to do so. When I came to Swansea in 2006, I only planned to stay for a year to do an MA in Translation with Language Technology. I already had a background in surveying (MSc in Geodesy and Cartography from Wroclaw University of Environmental and Life Sciences) and a passion for technical translation. While in the final stage of working on my MA dissertation in Swansea, Pius suggested I could combine the two in the research on surveying terminology. I am extremely grateful for his guidance and providing me with constructive criticism through the entire PhD project. I would also like to thank my second supervisor, Alison Williams, for painstakingly going through a very rough version of each chapter of my thesis and the entire manuscript, for correcting my clumsy formulations and indicating points that are ambiguous or not explicit enough to the reader.

I am also grateful to the Department of Languages, Translation and Media at Swansea University as I was able to benefit from their infrastructure and to Andrew Rothwell, the Head of the Department, for his comments and suggestions regarding the software aspects of terminology management. Thanks are also due to the Department of Computational Linguistics at Copenhagen Business School, and in particular to Hanne Erdman Thomsen, for giving me access to CAOS. In addition, I would like to thank my colleagues from the University with whom I had stimulating discussions over the years.

Finally, this research could not have been completed without the support of my family – my husband Karol, who helped me with many technical aspects of this research, our daughter Sara, who turned out to be a very cooperative baby, and my parents and parents-in-law.

INTRODUCTION

This book provides a comparison and analysis of surveying terminology in English and Polish. The purpose of the book is three-fold: firstly, to investigate how surveying terms are created and how they are named in English and Polish, secondly, to analyse concept systems of the two languages with respect to surveying terminology, and finally, to indicate the areas of surveying in which terminological and conceptual differences occur, the factors that trigger them and translation strategies which are used to solve them.

I have chosen surveying terminology for my research field as I have been educated and have professional experience in this field, which allowed me to discover that there are many terms which do not have equivalents in the target language, as well as concepts which occur only in one language. The name of the field itself is both intriguing and ambiguous. The field is commonly referred to as *surveying*, or most recently *geomatics*, in the Anglo-Saxon countries, while in the continental European tradition, it is called *geodesy* (*geodezja* in Polish). Arguably, the name *surveying* is too general because it indicates that anything can be surveyed, e.g. *literature survey*. I would personally opt for the name *land surveying*, which was used in the past. However, the paradigm changed in the Anglo-Saxon tradition and the current name of the field is *surveying*. Thus, to be consistent with the English conventions, I will refer to the field as *surveying* and will consider it as equivalent to *geodesy*.

Surveying terminology is an under-researched area and publications referring to surveying terminology in English and Polish are scarce. To my knowledge, there are four bilingual surveying dictionaries for English and Polish. Although they are called dictionaries, they are rather glossaries as they typically provide terms and their equivalents. The oldest one, the English-German-Polish dictionary by Tatarczyk (1991) includes 5,310 entries. A modified and expanded version of this dictionary was published on CD-ROM in 2005 (Tatarczyk, 2005). It includes 8,500 entries and covers various subfields of surveying and related disciplines, e.g. astronomy, civil engineering, physics, photography, photointerpretation, GPS, geology, computer science, mathematics, mining, remote sensing and optics. Apart from giving equivalent terms in the target language, the dictionary also provides gender and part of speech specification for German terms.

A second surveying dictionary is the Polish-English, English-Polish dictionary of terms from the fields of surveying, cartography and real estate by Downarowicz and Leśniok (2006). It includes 30,000 entries arranged in alphabetical order. It includes only terms and their equivalents in the target language.

The Internet dictionary of geomatics by Gaździcki (2005) is much more specialised than the other two dictionaries. However, at the same time it is quite limited as it covers only the part of the surveying domain that deals with state-of-the-art measurement technologies and computations. The dictionary is based on the printed version of *Leksykon geomatyczny - Lexicon of Geomatics* (Gaździcki, 2002). The lexicon was made available on-line in 2004 and updated in 2005. It includes both English and Polish entries. However, the Polish part of the dictionary contains more information, as, apart from equivalents, the dictionary also gives definitions for terms which are not available for English entries. Terms in the two languages are not the same; the Polish part includes more entries than the English part.

The English-Polish, Polish-English dictionary by Hycner and Szortyka (2005) is organised differently as it is divided into three parts. Part 1 is devoted to technical subfields of surveying such as geodetic surveying, cartography, GPS, photogrammetry and remote sensing, while part 2 is more legal-based, covering terms referring to cadastre. Part 3 includes names of the subfields of surveying and related disciplines. Each part is arranged in alphabetic order both in English and in Polish.

The above-mentioned dictionaries, although quite useful to surveyors, do not provide sufficient information for technical translators as they neglect problematic terms. Entries where conceptual mismatches occur are either simplified by providing a direct translation or omitted. For this reason, these dictionaries are of little help to those who are not experts in the field.

These dictionaries are the only publications on surveying terminology in English and Polish. There are no publications on terminological research in the surveying field. This type of research is required to enrich the content of surveying dictionaries and make them more useful for translators and technical writers, and also for specialists in the surveying field. Surveying terminology needs to be approached more systematically, preferably using a corpus-based study that provides evidence for documenting terms and concepts in addition to term-related and concept-related information.

My aim is to shed light on surveying terminology by combining terminological knowledge with expertise in the field in order to create two monolingual concept-oriented termbases with explicit information on concepts. These termbases will be used to identify conceptual mismatches and offer solutions for dealing with them. Data collected in these termbases will be analysed in the book to identify the differences between the terminological and conceptual systems in English and Polish.

The termbases include concepts mainly from three subfields of surveying: geodetic surveying, cartography and GPS, which have been selected out of ten subdomains of surveying with a reason. The first two fields are quite traditional and developed independently in different countries, whereas GPS is a relatively new field and has many recently-added terms. The methodology I developed for dealing with conceptual mismatches is extendible because I cover all types of problems that are expected in the field of surveying. It can be used to expand the research to the remaining seven subfields of surveying.

The book is structured into two parts. Part one is concerned with the compilation of termbases which document surveying terms and concepts (chapters one and two), and part two focuses on the analysis of terms and concepts in these termbases (chapters three, four and five).

After this introduction, chapter one describes how monolingual surveying termbases with translation equivalents were created in English and Polish. It discusses different approaches to terminology and terminological databases and their usefulness for creating surveying termbases (1.1). After that, it moves to a discussion of the problems with the name of the field and elaborates a classification of the field into a number of subfields (1.2). Next, selection of data categories for the surveying termbases is described (1.3), followed by a discussion of the termbase design and selection of the software (1.4).

Chapter two describes how surveying terms were collected. First, it discusses what approaches to terminology collection are available and selects those which are useful in this project (2.1). Then, it presents how the surveying corpora were designed and compiled (2.2) and how terms were extracted from them (2.3). It ends with a discussion of concordances (2.4).

Chapter three analyses how terms were created and how they were named. Starting from surveying termbases described in the course of previous chapters section (3.1) elaborates on linguistic processes which include word formation, and section (3.2) discusses term naming.

Chapter four concentrates on the meaning of terms. First, it examines various theories of meaning and their features (4.1). Then, it presents ontologies and their role in the representation of concept systems and discusses semantic relations between concepts (4.2). After that, conceptual mismatches in surveying termbases in English and Polish are described (4.3), followed by the presentation of translation strategies for dealing with them (4.4). Then, a classification of translation problems caused by conceptual mismatches is provided and solutions to selected conceptual mismatches are offered (4.5). The last section (4.6) concludes the whole chapter. Chapter five provides a summary of the research.

CHAPTER ONE

CREATING A TERMBASE FOR SURVEYING TERMINOLOGY

The aim of this chapter is to describe the working methods used in the creation of a termbase of surveying concepts. First, different approaches to terminology and terminological databases are discussed and their potential application for generating a surveying termbase is evaluated (1.1). Then, problems with the name of the field are discussed and a classification of the field into a number of subfields is developed (1.2). Next, an overview of data categories, their features and attributes is provided and individual decisions are made about which data categories to include in the surveying termbase. Choices for writing up terminological records are made at this stage (1.3). Finally, different software packages for terminology management are compared in order to select the one to be used in this project. These considerations are followed by a description of the design of the termbase in the selected software package (1.4).

1.1 Theoretical backgrounds

Terminology as a scientific discipline started to develop in the 1930s, when scholars from Russia, Austria and Czechoslovakia became aware of the proliferation of terms and the diversity of forms as well as relationships between terms and concepts (Cabré, 1999, p. 7). Modern terminological research was commenced by the engineer Eugen Wüster (1898-1977). His doctoral thesis (1931) was the first work on terminology. It outlined a new approach to terminology as the author did not focus on compilation of the specialised vocabulary or standardisation of existing terminology, but was interested in establishing principles for the creation of new terms (Pearson, 1998, p. 9).

His interest in the theory of terminology, known as the traditional approach to terminology, appeared more than thirty years later with the publication of *Die vier Dimensionen der Terminologiearbeit* 'The four dimensions of terminology work' (1969), in which he presented for the first time four aspects of terminology work: the special subject field, the languages, the purpose and the degree of abstraction. His overall approach to the theory of terminology, *Einführung in die Allgemeine Terminologielehre und terminologische Lexikographie* 'Introduction to the General Theory of Terminology and to terminological lexicography', was published posthumously in 1979 (Cabré, 1999, p. 225).

The traditional approach represented by Wüster and adopted by many later prominent terminologists was challenged by the corpus-based approach which evolved in the 1980s (Sager, 1990, p. 56). The traditional and corpus-based approaches constitute the mainstream approach to terminology, which is described in section 1.1.1. An alternative approach was developed by Rita Temmerman and is known as the sociocognitive theory of terminology. This approach is characterised in section 1.1.2. Apart from the traditional approach and the sociocognitive approach, there is a third approach to terminology, based on the FrameNet formalism. This approach evolved in parallel with the formation of the traditional approach and is presented in section 1.1.3.

1.1.1 The mainstream approach

The traditional approach to terminology is often referred to as the *general theory of terminology* (Pearson, 1998, p. 10). It is based on the theory developed by an Austrian engineer, Eugen Wüster. In his theory of terminology, Wüster (1979, p. 2) argued that the proper description of terms differs from the proper description of general language words. He suggested that work on terms differs from work on general language words in three respects. First, terminology work starts from the concept. Concepts exist independently of terms and any expression used to designate them should be considered in isolation from their labels or terms. Concepts are mental abstracts to which labels are assigned. The second distinction which Wüster makes refers to vocabulary. Wüster believes that terminologists are interested only in vocabulary and are not concerned with the theory of morphology or syntax. Traditional terminologists were not interested in examining terms in use as they only wanted to establish what they represented. The third distinction Wüster makes is about standardisation. Terminologists are concerned with imposing norms for the use of language. Their objective is to fix and standardise meaning in order to avoid confusion. This is achieved by creating a standardised collection of terms (Wüster, 1979, p. 2).

Wüster's approach to terminology is concept-oriented or onomasiological. It has been applied by many later terminologists. Wright (2001b, p. 579) claims that concept orientation is a recognised international standard for terminology databases. Warburton (2001, p. 687) adds that effective management of synonyms and equivalents requires them to be linked through the concept. Concept-oriented or terminographic collections consist of multiple records linked by a concept and are usually stored in a multidimensional structure, such as a database, with multiple access points to each record.

Wüster's approach was not ideal, as in many cases terminologists start their work not from concepts but from terms, which they find in the text or corpus of texts they examine. They first come across word forms and then they try to establish their meanings. The approach they take is thus corpus-based or semasiological.

The distinction between onomasiology and semasiology is a traditional one in continental structural semantics and in the Eastern European tradition of lexicographic research (Geeraerts, 2006, p. 37). It is quite hard to establish which of these two perspectives came first.

The onomasiological tradition seems to be older as it was used between the post-classical European written culture (c.800) and 1700 in various genres of text including non-alphabetical glosses, glossaries and dictionaries (Hüllen, 1999, p. 406). However, the term *onomasiology* was coined only in 1902 by the German linguist Adolf Zauner in his dissertation on body-part terminology in Romance languages (Grzega, 2002, p. 1021).

On the contrary, the term *semasiology* was created between 1822 and 1824 by a German scholar, Christian Karl Reising (1792-1829), who introduced an architecture of grammar that comprises an explicit semasiological component alongside traditional elements such as etymology and syntax (Schmitter, 2008, p. 575). Thus, in Reising's view, semasiology indicated the meaning of the word form.

Reising's ideas were popularised by his followers, Friedrich Haase (1808-1867) and Ferdinand Heerdegen (1845-1930). The term *semasiology* was also applied in England and the USA as well as in German-speaking countries. Its dominance lasted until it was replaced by *semantics*. The use of the term *semantics* by Charles R. Lanman (1850-1941), a scholar of Sanskrit at Harvard, in a lecture in the USA in 1894 marks the beginning of the gradual replacement of the term *semasiology*. In France, by contrast, the term *sémasiologie* was never in contention. The term which was established there was *sémantique,* introduced by Michael Bréal (1832-1915) in 1897 (Schmitter, 2008, p. 584). The term sémantique was adopted all over the world, while the popularity of the term semasiology decreased and it remained in use only in German-speaking countries. The term semasiology ceased to denote all aspects relating to the theory of meaning and was instead used with reference to a specific semantic perspective (Schmitter, 2008, p. 585), which starts from forms and looks for their meaning.

Two terms which received recognition in different parts of the world are *semiotics* and *semiology*. Both terms have the Greek word *semeion* 'sign' as their etymological source and refer to the study of signs. They have different histories, however. The term *semiotics* has an American origin as Charles Sanders Peirce (1839-1914), an American philosopher and logician, used several variants of the Greek word in his works: 'semeiotic', 'semeotic' and 'semiotic'. The diffusion of 'semiotics' as the currently accepted form began in the mid-1960s (Nuessel, 2006, p. 193) . The term *semiology* has a French origin as Ferdinand de Saussure, a Swiss linguist, used the French expression *sémiologie* to name the study that deals with social production of meaning from sign systems (Saussure, 1916/1969, p. 68). The term *semiology* travelled to the United States under the influence of Saussure's linguistic theories, while *semiotics* travelled to Europe and become the preferred designator of the field today (Nuessel, 2006, p. 193).

Referring back to *semasiology*, Karpova (2006, p. 709) claims that *semasiology* in its current sense was developed in the 1980s and covers various aspects of a word's semantic structure; from the simple correlation of a word and a concept, to the theory of reference and a description of a hierarchically organised structure of lexical entries. The development of computer technologies in the 1980s which facilitate the storage and management of large corpora in electronic form revolutionised the semasiological approach which became corpus-based. The semasiological perspective overshadowed the onomasiological one as most dictionaries use a universal alphabetical order for arranging entries (Hartmann, 2006, p. 669). Entries are derived from the corpus which provides other information on entries such as examples, grammatical data and semantic relations. The semasiological approach is applied by a number of terminologists including Jennifer Pearson, Juan C. Sager and Maria Teresa Cabré.

The corpus-based approach must not be confused with the corpus-driven approach. The fundamental distinction between these two methods was introduced by Tognini-Bonelli (1996). It was made on the basis of the purpose for which the corpus is being used. In the corpus-based approach, corpora are used mainly to "expound, test or exemplify theories and descriptions" (Tognini-Bonelli, 1996, p. 1). The corpus works as a repository. It is used, for example, to validate existing categories or different applications, to test a tagger or a parser (Tognini-Bonelli, 1996, p. 1). On the other hand, the corpus in the corpus-driven approach is more than a repository of examples to back pre-existing theories. It is also used to discover new facts in order to refine the hypothesis (Pearson, 1998, p. 49).

Initially, concept-oriented and corpus-based approaches were in conflict but in time they were integrated and nowadays they are often used together and complement each other. Van der Vliet (2006, p. 62) claims that a system of concepts, which is an intermediate structure between terms and domain knowledge, should be built by combining the top-down approach using domain knowledge and the bottom-up approach using a corpus. Terminologists using the top-down approach should be familiar with the subject field before they start structuring the field as it helps to build a concept system and establish relations between concepts. The top-down approach is concept-oriented (or onomasiological) as it starts from concepts and looks for their names. The candidate terms and collocations that can be linked to various concepts are extracted from a corpus of texts relevant for a particular domain by using the bottom-up approach.

The bottom-up approach starts from words and looks for their meanings. It relies on the corpus, which is examined to find the terms and to observe the way terms combine with other terms in compounds, collocations and sentences. Therefore, this approach is referred to as corpus-based or semasiological. In the corpus-based approach, the terms and their combinatoric properties are a basis for collecting and structuring domain knowledge (van der Vliet, 2006, p. 62).

Apart from strictly corpus-based or concept-oriented approaches, some terminologists, for example Wright (2001b, p. 552), suggest a third solution, which is based on the combination of both approaches. A corpus is the source of the information and a starting point for the research. A terminologist who applies this solution gets a list of terms as a result of the extraction process. These terms need definitions. While defining these terms and providing examples that document the use of the terms, the terminologist encounters new terms in definitions and examples and becomes aware of the concepts they represent. Then, he starts working in the opposite direction, from concepts to their meanings, looking at the concept structure of a particular domain and relations between concepts such as synonymy, hyperonymy, holonymy. The list of terms obtained in the extraction process is only a reference list. Terms from this list are usually entries in the termbase, but the number of terms in the termbase will be much higher than the number of terms on this list, as the terminologist aims to create a complete network of concepts and has to define all new concepts that occur in definitions and examples.

1.1.2 Sociocognitive approach

The sociocognitive approach to terminology was developed by Temmerman (2000). This approach relies on *socioterminology*, which is a relatively new trend in terminology that tries to get the study of terminology back to the study of real language usage (Boulanger, 1995, p. 197). Socioterminologists state that language is not suited for standardisation as it changes all the time. They also believe that social aspects, alongside cognitive ones, should be considered in terminological theory and practice. Temmerman reacts against the traditional onomasiological approach to terminology, against traditional definitions of concepts (consisting of an intensional or extensional definition) that reflect the position of the concept in a concept system, and against the univocity principle which states that there should be a one-to-one correspondence between a concept and a term. Temmerman strongly supports a corpus-based approach which is the starting point for her analyses. She uses a corpus of texts on the life sciences to carry out an empirical study of categorisation and lexicalisation processes. On the basis of her findings, she questions the validity of traditional approaches to terminology and suggests an alternative, which is inspired by the cognitive sciences. Temmerman (2000) bases her research and theory on such paradigms as hermeneutics (which is the main source of post-modernism represented by Derrida) and the cognitive approach in semantics, which both react against structuralism.

Temmerman (2000, p. 1-9) examines the term *concept* and possibilities for describing its meaning. She suggests that terminologists should start from *units of understanding* instead of *concepts*. She claims that *concepts* along with *categories* are two kinds of *units of understanding*. Only a few units of understanding do not have prototype structures and could therefore be named concepts (e.g. intron). The ones which have a prototype structure are categories (Temmerman, 2000, p. 43).

Temmerman (2000, p. 95) uses Idealised Cognitive Models (ICMs) to study units of understanding in the field of the life sciences. ICMs were discovered and described by Lakoff (1987) who, inspired by Fillmore's frame semantics, wrote a book entitled *Women, Fire and Dangerous Things. What Categories Reveal about the Mind.* The main message delivered by this book is that people organise their knowledge by means of structures called *idealised cognitive models* (ICMs) and category structures and prototypes are by-products of that organisation. Temmerman (2000) builds on Lakoff's theory and claims that ICMs may be perceived as conceptual prerequisites for understanding the meaning. ICMs are complex structures consisting of units of understanding. The intracategorial or internal structure (aspects, facets) of the units of understanding, as well as its intercategorial structure (the relationship of the given unit of understanding with other units of understanding with the same frame of mind), depend on the ICM within which a unit of understanding has been identified (Temmerman, 2000, p. 96).

Temmerman also believes that terminologists should replace traditional definitions by templates of meaning description. In traditional terminology, concepts are defined on the basis of necessary and sufficient characteristics (Temmerman, 2000, p. 226). In sociocognitive terminology meaning description can be presented via a template. There will be different types of templates and different modules of information within the template depending on whether the unit of understanding is an umbrella category, an entity or an activity. Different information modules can vary in information importance on a scale from 0 to 2 (from 0=irrelevant to 2=prominent). Temmerman also believes that synonymy and polysemy are functional in the process of understanding and that there does not need to be a one-to-one correspondence between the concept and the term. Polysemisation can occur due to a change in the world as new technology evolves or may be caused by a change in the understanding of the category.

Temmerman explores the link between the structure of understanding a category and the process of lexicalisation. She argues that two opposing forces are at work when categorisation takes place within a language community (Temmerman, 2000, p. 133). One of these forces is the urge for univocity and the other one is the urge for diversification (due to the evolution of categories and units of understanding over time, which results in polysemy and synonymy). She also analyses the mechanisms behind the urge for a new and better understanding of terms. Temmerman's hypothesis states that these mechanisms are related to, and inspired by, metaphorical reasoning (Temmerman, 2000, p. 158). Metaphor is a multidimensional phenomenon. Until the late 20th century it was used only to describe a literary figure of speech, but recently its use has widened to philosophy, psychology, linguistics and other cognitive sciences.

Lakoff and Johnson (1980) develop a new approach to the study of metaphor. They state that metaphor is present in everyday life, not just in language but also in thought and action. They also argue that human conceptual system is fundamentally metaphorical in nature. The central aspect of their argument is that metaphor is a kind of thinking or conceptualisation not limited to language. Language, however, helps to observe how metaphor works. They use a conceptual metaphor (or metaphorical conceptualisation), ARGUMENT IS WAR, to demonstrate how a concept can be metaphorical and can structure an everyday activity. The conceptual metaphor ARGUMENT IS WAR may be encountered in (1):

(1) Your claims are *indefensible.*
He *attacked every weak point* in my argument.
I *demolished* his argument.

The words in italics refer to war. Lakoff and Johnson (1980, p. 4) suggest that the examples given above provide evidence that our culture conceptualises arguments through the war metaphor and therefore the way in which people conduct arguments is conditioned by the way in which they conduct wars.

Lakoff and Johnson's view on metaphor, its role in everyday life and relation to language, was recognised and elaborated by many other linguists. For example, Knowles and Moon (2006, p. 4) claim that metaphor is a basic process in the formation of words and meanings. Concepts are lexicalised through metaphor. Many senses of polysemous words are metaphors of different types, e.g. *jewel* is a metaphor for something valuable, *fox* is a metaphor for a cunning, wily person.

Temmerman carried out a study of metaphorical lexicalisations or part of metaphorical ICMs (m-ICMs) in the language of the life sciences. The aim of her study is to prove the non-arbitrariness of the sign in the life sciences, in the sense that metaphoric models result in lexicalisations. She wants to prove that terms are motivated. It is a very interesting approach as it opposes the Saussurian principle of the arbitrary character of linguistic signs, which says there are no motivated links between the signifier and the signified (Saussure, 1916/1969, p. 66). The main outcome of Temmerman's study is a description of metaphorical models that explain the phenomenon of metaphor at the lexeme, category and domain levels.

Having analysed features of traditional principles and new propositions, Temmerman (2000, p. 223) develops the following set of alternative principles to traditional terminology:

1. Sociocognitive terminology starts from units of understanding - most of which have a prototype structure.
2. As understanding is a structured event, a unit of understanding has an intercategorial and intracategorial structure and it functions in cognitive models.
3. Depending on the type of unit of understanding and on the level and type of specialisation of sender and receiver in communication, what is more essential or less essential information will vary.

4. Synonymy and polysemy are functional in the progress of understanding and therefore need to be described.
5. Units of understanding are constantly evolving; cognitive models (e.g. metaphorical ICMs) play a role in the development of new ideas which implies that terms are motivated.
6. Sociocognitive terminology supports a combined semasiological and onomasiological approach to terminography. It takes into consideration the role of metaphorical idealised cognitive models and synonymy and polysemy in the process of understanding. It replaces traditional definitions with templates for meaning description. The distinction between concepts and categories is central within sociocognitive terminology. Temmerman's approach questions the validity of traditional terminology as, in practice, terminologists start their work from concepts rather than from a list of terms.

Temmerman's approach, in a slightly modified version called termontography, is currently used at the Erasmushogeschool in Brussels, where she is a lecturer (Thelen & Steurs, 2010, p. 260). Termontography is a multidisciplinary approach in which theories and methods for multilingual terminological analysis of sociocognitive theory are combined with methods and guidelines for ontology engineering (Centrum voor Vaktaal en Communicatie, 2009). In this approach, a clear distinction is made between conceptual modelling at a language-independent level and language-specific analysis of units of understanding. A key view in termontography is that knowledge analysis should precede the methodological processes which are generally conceived as the starting-points in terminography, i.e. the compilation of a domain-specific corpus of texts and the understanding and analysis of the categories that occur in a certain domain. This view is supported by the fact that the aim of terminological databases is to represent in natural language those items of knowledge or *units of understanding* which are regarded as relevant to specific purposes, applications or groups of users. In termontography, the units of understanding and their intercategorial relations are structured in a common knowledge base or categorisation framework. On the one hand, this framework supports the information-gathering phase during which a corpus is developed. On the other hand, it allows terminographers to establish specific extraction criteria to define what should be considered a 'term'. Moreover, pre-defined knowledge also influences the terminographer's working method and the software tools that will be employed to support that working method.

1.1.3 FrameNet

FrameNet is a computational lexicographic project whose purpose is to represent information about the semantic and syntactic properties of English words and encode this information in a database (FrameNet, 2008). This information is extracted from large electronic corpora using manual and electronic techniques. Although FrameNet was not developed for terms, the database and its construction technique may be very useful in describing them (van der Vliet, 2006). FrameNet is based on the theory of Frame Semantics, whose central idea is that word meanings must be described in relation to semantic frames.

The linguistic basis of FrameNet is Fillmore's theory of Case Grammar (Fillmore, 1968). FrameNet itself was developed in two stages: FrameNet I and FrameNet II (FrameNet, 2008). The FrameNet database consists of a lexical database and an annotation database. The lexical database contains entities of the following types:

- frames: named data structures which are used to represent a concept in a domain, e.g. *survey;*
- frame elements (FEs): kinds of entities that can participate in the frame, e.g. FEs of the frame *survey* are: *contractor*, *object*, *purpose;*
- lexemes: sets of forms taken by a single word (e.g. the English verb lexeme *measure* has four word forms: *measure*, *measures, measured*, *measuring*);
- lemmas: particular forms of the lexeme that are chosen by convention, e.g. *measure* is the lemma of the verb lexeme *measure.* The lemma is the canonical form of a lexeme and in lexicography it is usually also the citation form or headword by which it is indexed;
- lexical units (LUs): associations between lemmas (units of form) and frames (units of meaning). LUs correspond to dictionary senses.

The following types of relations between frames and frame elements exist in FrameNet:

1. *Frame Inheritance*. Frame Inheritance is a relation between frames which depends on the fact that a more specific frame inherits from a more general frame, e.g. the *Surveying_tools* frame evoked by the verbs *measure*, *set out* and *level* inherits from the more general

Surveying frame that has the same FEs, e.g. *method*, *participants*, *accuracy* and *purpose*.
2. *The Subframe Relation.* Subframes describe subevents which are part of complex events described in terms of frames, e.g. the *cadastre* frame is a complex event and it consists of such subframes as: *ownership*, *location*, etc.
3. *The 'Uses' Relation.* This relation is similar to inheritance but less strictly defined. In this relation, frame *A* uses frame *B* and frame *A* need not have a FE corresponding to each FE of frame *B*, e.g. the *Plan* frame uses the *Map* frame, but whereas the *Map* frame has a FE *projection*, the *Plan* frame does not.
4. *The 'See also' relation.* The 'see also' relation is a pointer from one or more frames to another frame whose definition contains a discussion of differences among the frames in the group. For example, the two uses of words like *load* in *She loaded the wagon with hay* and *She loaded the hay onto the wagon* are treated in the two frames *Filling* and *Placing* respectively (Baker, Fillmore & Cronin, 2003, p. 287). The detailed discussion of differences between these two frames is located in the definition of *Filling* and there is therefore a reference, *'see also'* relation, from *Filling* to *Placing*. There is, however, no systematic solution to the problem of relating these two words. It is done on an ad hoc basis.

The annotation database contains sentences and their annotation. For each lemma, there is a set of annotation layers for frame elements, phrase types, grammatical functions, etc. The lemma, along with its layers, is represented by an entry in the Annotation Set table, which links a sentence, a subcorpus and a LU.

The annotation process depends on labelling sentences using labels that indicate various semantic and syntactic properties. It is an interactive and semi-automatic process. The Semantic Annotator program presents the sentences one at a time and provides menus that identify the available syntactic phrase types, the grammatical function of the constituents and the frame element names for the target frame. The operator's role is to identify the sentence constituents corresponding to frame elements, and to tag them according to their grammatical function (subject, object, oblique), their phrase type (NP, PP, VP, etc.) and their frame element (Lowe, Baker & Fillmore, 1997).

The lexical database may include semantic types, which were mentioned above and which are used to recognise some aspects of the meaning, e.g. positive and negative connotations. It can also contain notes which are used to record questions and problems that arise when the lexical or annotation database are created.

However, certain types of lexical information are not covered in the FrameNet database compared to dictionaries. The FrameNet database does not include phonological, morphological or etymological information about the words in the database. It does not provide information about lexical relations, e.g. synonymy, antonymy or hyponymy, either. Finally, it does not offer any statistical data about the frequency of occurrence of syntactic patterns or about LUs.

A FrameNet database is implemented as a MySQL database. Three types of interface have been developed for this database. One is for browsing, the second is for searching and the third is for updating the database. The FrameNet database is much more informative and systematic than a paper dictionary or even an average electronic dictionary. There is no fixed limit to the number of examples. It also has room for information on combinatorics, e.g. which preposition follows a certain verb and which verb to use in combination with a noun. The use of frames facilitates presenting this information as systematically as possible. A FrameNet database may be used for the production of dictionaries and in various projects in the domain of language technology.

Examples where FrameNet formalism is used include the development of multi-lingual lexica, word sense disambiguation and machine translation. The formalism is also used in Natural Language Processing (NLP) systems that perform question answering, information retrieval and automatic semantic parsing (Baker & Sato, 2003). FrameNets have been built for other languages such as German, Spanish, Japanese or even Polish. While the FrameNets in the first three languages are very well developed and have their entries linked with English-based FrameNet lexical entries to arrive at a contrastive FrameNet lexicon, the Polish FrameNet is only a small project whose aim is to describe a subset of about 200 Polish verbs using Frame Semantics. The Polish project is described by Zawisławska (2010).

One of the most practical applications of FrameNet is the *Kicktionary* created in 2006 which is a domain-specific trilingual (English, German, and French) lexical resource of the language of Soccer, created by the FrameNet visitor, Thomas Schmidt from Germany. Kicktionary relies on Frame Semantics and uses semantic relations in WordNet style as an additional layer of structure (FrameNet, 2008). The lexicon currently contains around 1,900 lexical units which are organised in 104 frames and 16 scenarios. Each LU is illustrated by a number of examples from a multilingual corpus of football match reports from the UEFA website (Schmidt, n.d).

The idea of FrameNet databases is also applied in the *Referentiebestand Nederlands (RBN)* which is a multi-purpose lexical database of Dutch (van der Vliet, 2007). The case for using FrameNet in a terminological context was made by van der Vliet (2006, p. 57) who suggests that the FrameNet approach provides a way of describing the semantics of terms in a frame-based way by linking them to concepts. The use of frames has many advantages. It leads to the establishment of an explicit link between terms and the domain knowledge, increases consistency by the systematic description of domain knowledge and provides the possibility of basing a definition on a selection of the represented knowledge, which leads to more flexible definitions (van der Vliet, 2006, p. 61). A database with frame-based description has better retrieval possibilities. It can be entered through terms but it additionally benefits from the slots and fillers which are used to search for individual terms or sets of terms.

1.1.4 Conclusions

When the mainstream approach, the sociogonitive approach and FrameNet are compared, it may be noticed that they differ quite significantly. The mainstream approach is a combination of the onomasiological and corpus-based approaches. It originated as a traditional onomasiological approach but after the expansion of the use of computers in the 1980s it was followed by the corpus-based approach with which it was later integrated (ten Hacken, 2010b).

The insights from the sociocognitive approach are more radical. Temmerman (2000) uses the corpus-based approach as a tool to obtain textual evidence confirming the limitations of the traditional approach. She reacts critically against the onomasiological perspective, traditional definitions of terms and univocity. She suggests that the semasiological approach and templates for units of meanings would be much more efficient. She also recognises polysemy and synonymy as useful in terminology.

FrameNet is a contribution to terminology of a different type from the traditional corpus-based or sociocognitive approaches. It includes guidelines on how to build a concept system based on frames and presents the advantages which a database with a frame-based description of domain knowledge offers. The FrameNet approach is lexicographic but it can also be applied in terminology.

1.2 Surveying

The domain of surveying is very challenging for terminologists because of discrepancies in the naming of the field and its complex structure. The name of the field has evolved over time from *land surveying* to *surveying* and *geomatics*. A lack of international agreement as to how the field should be named and different naming conventions, both in continental Europe and the Anglo-Saxon countries, has increased the number of different labels used for the field. Furthermore, the field of surveying can be divided into a number of subdomains, e.g. cadastral surveying, plane surveying. Different classification schemes provide an overview of the field, but they often differ significantly in the subfields they specify.

In this section I will look at differences in the naming of the field and analyse a range of labels under which the field is known (1.2.1). I also compare and contrast classification schemes in surveying (1.2.2) and I attempt to develop a uniform classification system for this field (1.2.3) based on the analysis in the first two subsections.

1.2.1 Differences in the naming of the field

The term *surveying* is quite ambiguous as the field is named, perceived and understood differently in different countries, with notable differences existing between the continental European tradition and the Anglo-Saxon tradition. The field is named *surveying* (formerly *land surveying*) in the Anglo-Saxon tradition (Ghilani & Wolf, 2008, p. 1) and *geodesy* in the European tradition (Hycner & Dobrowolska-Wesołowska, 2008, p. 16). I will try to provide definitions and examples of usage of these terms in order to find differences in the meaning between them.

I will now look at how these terms are described in a number of textbooks on surveying, encyclopaedias and standards. In the case of *surveying*, I have an article by Lyman and Wright (2009) in the *Encyclopaedia Britannica* and a surveying textbook by Bannister et al. (1998). According to Lyman and Wright (2009), *surveying* is a method of making relatively large-scale, accurate measurements of the Earth's surface. Its principal modern uses are in the fields of transportation, building, land use, and communications. Surveying is divided into plane surveying, which deals with mapping small areas, such as a building site or a parcel, and geodetic surveying, which focuses on mapping large areas of the globe, e.g. the territory of a country. The definition of surveying by Bannister et al. (1998) and the division of surveying into geodetic and plane surveying are consistent with Lyman and Wright's view on this field.

Encyclopaedia Britannica (2009) defines *geodesy* as the scientific discipline concerned with the precise figure of the Earth and its mathematical description. Until the advent of satellites, all geodetic work was based on land surveys made by triangulation methods employing a geodesic coordinate system. It is now possible to use satellites along with the land-based system to refine knowledge of the Earth's shape and dimensions. Therefore, a new branch of geodesy, often called *satellite geodesy* evolved. The term *geodesy* is also recorded in Bannister et al. (1998), and is understood as the study of the size and shape of the Earth and its gravity field. It gives rise to the name *geodetic surveying*, which is a branch of surveying. This information is consistent with what Lyman and Wright (2009) state. Hence, I notice a paradigm change from *land surveying* as the name of the field that deals with boundary measurements to *geodesy* which deals with the shape and the gravity of the Earth.

Although, from the discussion provided so far, one may assume that *geodesy* and *surveying* are equivalent terms, there a number of examples in literature that contradict this. For example, in *ISO/TR 19122: 2004* standard there are statements that suggest that geodesy and surveying are two different domains: "In October 2001, the 10-semester program of Surveying and Geodesy at the Graz University of Technology has been replaced by a 6-semester bachelor program "Bakkalaureat Geomatics Engineering" and a subsequent master program "Magisterstudium Geomatics Science" (ISO, 2004, p. 45) and "There are 3 courses of Geographic Engineering at university level. These courses encompass the fields of Geodesy, Photogrammetry, Remote Sensing, Surveying and GIS". (ISO, 2004, p. 50). These examples indicate that *surveying* and *geodesy* do not refer to one concept in Austria but are labels for two different concepts. I will now analyse how the field is perceived and named in Poland.

Polish surveyors and scientists note the discrepancy in naming the field. Hycner and Dobrowolska-Wesołowska (2008, p. 16) stress that there are many terms in geodetic terminology that lack Polish equivalents and the terms *surveying* and *geodesy* seem to be the best examples. In their opinion, surveying or plane surveying deals with the determination of the relative spatial location of points on or near the surface of the Earth. It depends on measuring the slope and vertical and horizontal distances between objects, measuring angles between lines, determining the direction of lines and establishing point locations by predetermined angular and linear measurements (Hycner & Dobrowolska-Wesołowska, 2008, p. 20). The term is translated into Polish as *miernictwo* 'measurement', i.e. surveying or *geodezja na płaszczyźnie* 'geodesy on a plane', i.e. plane surveying. The term *surveying*, in Hycner and Dobrowolska-Wesołowska's view has a broad and confusing connotation as it describes both procedures and studies. It becomes more definite when used with other terms, e.g. *cadastral*, *property*, and *boundary*. The term *geodetic surveying* or *geodesy* describes the type of surveying that takes into account the true shape of the Earth. Geodetic surveys are of high precision and extend over large areas (Hycner & Dobrowolska-Wesołowska, 2008, p. 24). *Geodesy* is translated into Polish as *geodezja* 'geodesy' or *geodezja wyższa* 'higher geodesy'. This view of the meaning of the terms *surveying* and *geodesy* is confirmed by equivalents given in Polish-English, English-Polish geodetic dictionaries. The term *surveying* is translated as *miernictwo* 'measurements', i.e. surveying, and the Polish equivalent for *geodesy* is *geodezja wyższa* 'higher geodesy' in the dictionary by Downarowicz and Leśniok (2006). The same equivalents are provided in the English-German-Polish electronic dictionary of surveying compiled by Tatarczyk (2005).

If I follow Hycner and Dobrowolska-Wesołowska's reasoning (2008) when I look for the equivalents of Polish terms in English, I should translate *geodezja* 'geodesy' as *land surveying* because it incorporates both *miernictwo* 'measurement', i.e. *plane surveying* and *geodezja wyższa* 'higher geodesy', i.e. geodetic surveying.

The point Hycner and Dobrowolska-Wesołowska (2008) make on the concept structure of the field is in line with what ISO/TR 19122: 2004 states. They indicate that in countries such as Germany and Poland terms corresponding to *geodesy* and *surveying* are used to name the subfields of the field in question and therefore, when translating the Polish name of the field into English the distinction should be made between the holonym and meronym. In order to validate this claim, I will look at how the field is referred to in the education systems in Poland and in the United Kingdom.

In Poland, five major public universities offer undergraduate courses in Geodesy and Cartography: AGH (Akademia Górniczo-Hutnicza 'Academy of Mining and Metallurgy') University of Science and Technology in Kraków, University of Agriculture in Kraków, Wrocław University of Environmental and Life Sciences, Warsaw University of Technology and University of Warmia and Masuria in Olsztyn. Undergraduate geodetic courses are also available at a number of private, recently-founded universities. These courses offer a wide range of modules that include basic linear and angular measurements (taught in the first year under the name Geodezja I 'Geodesy I'), electromagnetic distance measurements and tachometric measurements (taught in the second year of undergraduate studies as Geodezja II 'Geodesy II') and satellite measurements (taught in the third year as Geodezja wyższa 'Higher geodesy'). Students who decide to pursue postgraduate studies can select a narrower specialisation such as Real Estate Management, Civil Engineering, Surveying, Geomatics or Geographical Information Systems.

Education of students in the field of surveying in the United Kingdom seems to be quite different. The UCAS (2011) search for *geodesy* gives no results, but there are numerous examples of universities listed when a *surveying* course is searched for. The evidence shows that surveying seems to be a component of various academic degrees such as Building Surveying at the University of Salford or Quantity Surveying at Swansea University. Not all the hits for surveying are relevant to geodesy, however. Southampton University offers a course in Yacht Production and Surveying which does not have anything in common with the type of surveying that is understood as geodesy. In fact, surveying in this course encompasses the Yacht and keel surveying module in year 2 and Yacht hull surveying in year 3. This example confirms that the denotation of surveying in English is quite broad and, to some extent, justifies concerns Polish surveyors have (Hycner & Dobrowolska-Wesołowska, 2008, p. 24) when translating the term *surveying* as *geodezja* 'geodesy'.

To sum up this discussion, there are differences in the perception of the field in the Anglo-Saxon countries and in continental Europe. The term *geodesy* is common in continental Europe, while the Anglo-Saxon countries use the term *surveying* as the preferred term and make a distinction between geodetic surveying and plane surveying. The former is concerned with measuring large areas of the country and may involve satellite surveys as it takes the curvature of the Earth into account, while the latter focuses on measurements of small parts of the country and assumes that the Earth is flat. The meaning of the term *plane surveying* is wider than of *land surveying* as it deals with all types of measurements of small areas of land, while the main task of land surveying is to measure boundaries.

When the definitions of the terms *geodesy* and *surveying* are compared, it may be noticed that they seem to be equivalents if English is taken as a starting point for the analysis. However, if this concept is looked at from the perspective of Polish scientists and surveyors, it may be learnt that they translate *surveying* as *miernictwo* 'measurements' and *geodezja na płaszczyźnie* 'geodesy on the plane' and *geodesy* as *geodezja wyższa* 'higher geodesy'. They tend to translate the name of the whole field as *land surveying*. This last instance is inconsistent with English, where the concept of land surveying covers only a small part of the field concerned with measurements of boundary lines. Therefore, from this point on in my thesis, I will refer to the field in question as *surveying*.

1.2.2 Classification systems

There is no standardised classification of surveying fields. The two main sources that include some sort of categorisation of the field are general systems for classification of knowledge and textbooks. In the discussion below I will examine the Universal Decimal Classification, abbreviated as UDC (UDC Consortium, 2010) and the Encyclopaedia Britannica (1986) classification, which are general classification systems, and classification in the surveying textbook by Bannister et al. (1998).

The Universal Decimal Classification (UDC)

The UDC is a system of library classification developed by the Belgian bibliographers Paul Otlet (1868-1944) and Henri la Fontaine (1854-1943) at the end of the 19th century (UDC Consortium, 2010). It is based on the Dewey Decimal Classification (DDC), a proprietary system of library classification developed by Melvil Dewey (1851-1931) in 1876. The DDC organises all knowledge into ten classes, which are further subdivided into divisions and sections. Each main class has ten divisions and each division has ten sections. The UDC uses Hindu-Arabic numerals and is based on the decimal system. Every number is thought of as a decimal fraction with the initial decimal point omitted, which determines filing order. For ease of reading, a UDC identifier is usually punctuated after every third digit. The advantage of this system is that it is indefinitely extensible, and when new subdivisions are introduced they need not disturb the existing allocation of numbers.

This system has 10 main numbers, from 0 to 9, which relate to the following fields:

0	Generalities
1	Philosophy, Psychology
2	Religion, Theology
3	Social Sciences
4	Vacant
5	Natural Sciences
6	Technology
7	The Arts
8	Languages, Linguistics, Literature
9	Geography, Biography, History

In the UDC, surveying is included under point 5 (*Natural Sciences*), sub-point 52 (*Astronomy. Astrophysics. Space research. Geodesy*), as section 528 (*Geodesy. Surveying. Photogrammetry. Cartography*). The UDC is described thoroughly in British Standards. Each of the standards is devoted to a different UDC subject. BS 1000 [52]: 1977 covers UDC 52 *Astronomy. Astrophysics. Space research. Geodesy.* The table of contents of BS 1000 [52]: 1977 standard demonstrates that it contains further auxiliary subdivisions into the following categories:

520	Instrumentation and techniques (astronomy and astrophysics)
521	Theoretical astronomy. Celestial mechanics. Fundamental astronomy. Theory of dynamical and positional astronomy
523	The Solar System
524	Stars and stellar systems. The Universe
527	Navigational Astronomy
528	Geodesy. Surveying. Photogrammetry. Remote Sensing. Cartography
529	Chronology. Calendar. Determination of time

It is important to note the systematic use of vacant categories in UDC in addition to the vacant field 4. While the vacant field is designed for a completely new branch of knowledge, vacant categories such as 522, 525 and 526 within the class 52 leave space to the emergence of new categories or reclassification of existing categories when they become too extensive.

The list of related subjects in the initial part BS 1000 [52]: 1977 standard points out fields that are interrelated with point 52 and it suggests such subjects as the following:

333	Land and landed property
622	Mining. Mineral dressing
623	Military engineering
624	Civil and structural engineering in general
625	Civil engineering of land transport
744	Linear, geometric, technical drawing, etc.
912	Nonliterary, nontextual representations of a region

Most of the issues of surveying are covered by class 528. Points 521 to 527 and point 529 also describe subjects that are applied in surveying, e.g. astronomy and navigation, but they are not core to surveying. Although people involved in surveying benefit from information included in these points, they do not have to know how this knowledge is used in the original context. Points listed as related subjects, e.g. point 6, which is Technology, also cover topics related to surveying. However, examination of points 624 and 625 reveals that issues of Civil Engineering are approached from a different perspective in these points. These points deal more with building and construction than with surveying. They contain information on how to dig foundations, build tunnels, conduct excavations, construct bridge tunnels, etc. Point 528 of the UDC for surveying is a good starting point for determining subfields of surveying. The principal divisions in this point listed below suggest possible subfields of surveying:

528.1	Theory of errors and adjustment (geodetic and photogrammetric applications)
528.2	Figure of the Earth. Earth measurement. Mathematical geodesy. Physical geodesy. Astronomical geodesy
528.3	Geodetic surveying
528.4	Field and land surveying. Cadastral survey. Topography. Engineering survey. Special fields surveying
528.5	Geodetic instruments and equipment
528.6	Photogrammetry (aerial and terrestrial). Methods and instruments
528.7	Cartography. Mapping (textual publications)

At first sight, points 528.3 and 528.4 contain overlapping information because *Geodetic surveying* covers all types of measurement including *Cadastral surveys* and *Engineering surveys*. In order to identify the differences between the two points all sub-points of points 528.3 and 528.4 need to be considered as it enables one to learn how the classification was developed for these points.

528.3	Geodetic Surveying
528.31	Outline and structure of the geodetic survey. Orientation. Point of the origin. Laplace stations. Datum plane
528.32	Base-line measurements
528.33	Trigonometric networks
528.34	Special methods of trigonometric determination of stations
528.341	Flare triangulation
528.342	Solar eclipse method
528.343	Star occultation method
528.344	Stellar triangulation
528.35	Trilateration
528.37	Kinds of elevation and principles of levelling
528.38	Figuration and observation of level networks
528.381	Primary level networks. First-order level network
528.385	Filling-in of level networks. Level networks of second order, of lower orders
528.388	Benchmarks. Levelling stations
528.389	Special applications of levelling and depth measurement
528.4	Field and land surveying. Cadastral survey. Topography. Engineering survey. Special fields surveying
528.41	Local geodetic networks
528.42	Topographic surveying
528.44	Cadastral survey
528.441	Cadastral resurvey
528.442	Cadastral revision
528.443	Gradual re-establishment of the cadastral survey data
528.45	Urban surveying
528.46	Surveying for soil improvement
528.47	Hydrographic and coast surveying
528.48	Engineering survey. Special fields of surveying
528.481	Observations of local displacements of the ground
528.482	Measurement of deformation
528.484	Connection with underground survey

528.486	Staking-out (setting-out) operations for the transfer of a construction plan into the terrain. Tracing
528.489	Special fields of surveying

The above detailed classification of points 528.3 and 528.4 shows the difference between *Geodetic surveys* and other types of surveys specified in point 528.4, e.g. *Cadastral and Engineering surveys*. Point 528.3 illustrates basic techniques, while point 528.4 is concerned with their application. *Geodetic surveys* are the most basic surveys used in surveying and they provide the basis for other types of surveys. *Geodetic surveys* do not require any specialist equipment and can be performed irrespective of atmospheric conditions. The results of these surveys may be processed with the simplest geodetic software or even with a calculator. Geodetic surveys are taught to first-year students of surveying. This classification also includes 528.45 and 528.47 as special types of surveying rather than separate disciplines. Wright (1982) lists *Simple field surveying* which is equivalent to 528.3 as one of the disciplines in surveying but does not mention 528.45 and 528.47 in his classification. His book is directed mainly at first-year students. The UDC proves to be a very useful source of information on the classification of the field of surveying because it indicates how it is structured and indicates domains that are related to this field. It also specifies measurement methods in the field and their application.

The New Encyclopaedia Britannica classification

Another system of classification is used in the *The New Encyclopaedia Britannica* (to which I will refer as *EB*). The encyclopaedia has an interesting three-part structure. It consists of 12 volumes of *Micropaedia*, 17 volumes of *Macropaedia*, a single volume of *Propaedia* and two volumes of indexes. The *Macropaedia* consists of 681 long articles, whilst in the *Micropaedia* there are tens of thousands of short articles. The *Propaedia* gives a hierarchical outline of human knowledge and it includes a systematic Table of Contents which gives readers an overview of the classification system applied in *EB*. According to this Table of Contents, information in *The New Encyclopaedia Britannica* is divided into ten parts that refer to the following branches of knowledge:

Part One	Matter and Energy
Part Two	The Earth

Part Three	Life on Earth
Part Four	Human Life
Part Five	Human Society
Part Six	Art
Part Seven	Technology
Part Eight	Religion
Part Nine	The History of Mankind
Part Ten	The Branches of Knowledge

Each of these parts is followed by a few pages of introduction which explain what the part will contain and further subdivisions into sub-parts which are called *Divisions* in the *Britannica.* Under each Division, the component Sections are listed.

While Parts 1-9 present classification of knowledge, Part 10 is concerned with the classification of the study of knowledge. It mainly covers the disciplines or branches of knowledge themselves, e.g. Division III (*Sciences*) includes such disciplines as *The Earth Sciences*, *The Biological Sciences* and *Technological Sciences*, while Division II (*Mathematics*) covers *History and Foundations of Mathematics*, *Branches of Mathematics* and *Applications of Mathematics*.

The issues regarding Surveying are covered in Part Seven, Division II (*Elements of Technology*), Section 723 *Technology of Measurement, Observation, and Control*. This Section is further divided into sub-sections A to G, and under these subsections further points are listed. Sub-section G "Major systems of measurement and observation", seems to be the most relevant for the purpose of surveying because it covers such branches as "Surveying, Mapping and Cartography", "Astronomical observations and Navigational techniques and devices". Whenever necessary, points include references to other sections where extra information on the particular subject may be found. Each division of the *Encyclopaedia Britannica* in the *Propaedia* is followed by a list of suggested further reading in the *Macropaedia* and selected entries of reference information in the *Micropaedia*.

When the *UDC* and *EB* classifications are compared, a number of similar features may be recognised. Both classification systems consist of ten parts. The scope of knowledge they cover is the same in both cases, only the way of organizing the information differs, e.g. the *UDC* has a part called *Geography, Biography, History*, and the *EB* splits the same data into parts: *Life on Earth*, *Human Life*, *History of Mankind* and *Branches of Knowledge*. The *EB* seems to be too general in terms of the technical sciences. Its categorisation is quite superficial in comparison to the *UDC*.

It gives just the main names of subject fields, e.g. *Mapping and Cartography*, while the *UDC* makes a distinction between *Thematic Cartography*, *Theoretical Cartography*, *Practical Cartography*, etc. and then specifies aspects of each of these fields, e.g. *Generalization, Scales and Sheet Size* in the case of *Theoretical Cartography*. The *UDC* system is more efficient for surveying classification than the classification system of the *EB*. The classification in *the EB* is closed as it cannot be upgraded on a regular basis. The last printed edition of the *EB* started to be published in 1985 and is ongoing. The encyclopaedia is an extensive source of knowledge which not only includes classification of the knowledge, but also provides extensive semantic information on its entries. The huge effort and costs involved in the publication of encyclopaedia, means that new editions appear on average only every 10-15 years. In contrast, the *UDC* must be open because it is designed for continuous use. Its system can be extended but cannot be modified. It is used worldwide in bibliographic services, documentation centres and libraries, so the alteration of its structure would involve restructuring systems of ordering information in all the institutions that use it.

Classification in surveying textbooks

Another source of classification is found in surveying textbooks. The example chosen is *Surveying*, by Bannister et al. (1998), a standard surveying textbook and one which is widely read. Bannister et al. (1998, p. 3) suggest classifying surveys by purpose, as follows:

A Topographic surveys
B Engineering surveys
C Cadastral surveys
D Geographic information systems

This classification covers the following aspects of surveying which are specified in the table of contents:

1 Tape and offset surveying
2 Levelling
3 The theodolite and its use
4 Electromagnetic distance measurement,
5 Satellite positioning systems
6 Survey methods
7 Analysis and adjustment of measurements

8 Areas and volumes
9 Setting out
10 Curve ranging
11 Hydrographic surveying
12 Photogrammetry

As may be noticed, the classification system in Bannister's book is quite general and classifies surveys on the basis of purpose. Therefore, the reader may have the impression that it is not complete because it covers fewer fields than the UDC. Points A to D cover aspects from the table of contents of the Surveying textbook, e.g. Engineering survey includes: setting out and curve ranging which are the measuring techniques used in engineering surveys. Some aspects from the table of contents are in fact the names of fields in the UDC, e.g. Photogrammetry. Some techniques of the field of geodetic surveying are specified in the table of contents in Bannister et al. (1998) but the name of the field is not listed in this classification. In

Table 1-1 below, the fields of surveying specified by Bannister et al. (1998) have been amalgamated with the aspects of surveying in such a way that aspects have been assigned to fields. Most aspects could not be assigned to any field and therefore the UDC classification was used for comparison and filling the gaps. Juxtaposition of surveying aspects and surveying fields enables Bannister's classification to be extended with new fields which are specified in the UDC but do not occur in Bannister et al. (1998).

It is interesting to note that some fields, e.g. *Engineering surveys* (B) have several aspects, while other fields, e.g. *Geographic information systems* (D) do not have any aspects specified. Only two aspects *Setting out* (9) and *Curve ranging* (10) could be assigned to a field in the classification by Bannister et al. (1998). The affiliation of all other aspects required consulting the *UDC*. It turned out that the field of *Geodetic surveying* specified in the *UDC* includes as many as 6 aspects of surveying specified by Bannister et al. (1998) and that some aspects specified by Banister et al. (1998), e.g. *Photogrammetry,* are fields in the *UDC*. Some aspects and fields have slightly different names but they indicate the same entity, e.g. *Analysis and adjustment of errors* in Bannister et al. (1998) and *Theory of errors and adjustment* in the *UDC* or *Hydrographic surveying* in Bannister et al. and *Hydrographic and coast surveying in* the *UDC*. The aspect *Satellite positioning systems* is very specific, while the corresponding field in the *UDC* is more general and includes issues that have been already covered by other fields of surveying, e.g. *Mathematical geodesy* is a part of *Geodetic surveying* as calculations follow the measurements taken using various instruments, e.g. theodolites, levels, etc. Thus, I will consider aspects 5, 7, 11 and 12 to be subfields of surveying.

Table 1-1 Extension of the classification system found in Bannister et al. (1998)

Fields of surveying	Aspects covered by a particular field
Topographic surveys (A)	
Engineering surveys (B)	Setting out (9), Curve ranging (10)
Cadastral surveys (C)	
Geographic information systems (D)	
Geodetic surveying (528.3 in the UDC)	Tape and offset surveying (1), Levelling (2), The theodolite and its use (3), Electromagnetic distance measurement (4), Survey methods(6), Areas and volumes (8)
Theory of errors and adjustment (528.1 in the UDC)	Analysis and adjustment of errors (7)
Photogrammetry (aerial and terrestrial). Methods and instruments (528.6 in the UDC)	Photogrammetry (12)
Figure of the Earth. Earth measurement. Mathematical geodesy. Physical geodesy. Astronomical geodesy (528.2 in the UDC)	Satellite positioning systems (5)
Special fields of surveying (528.489 in the UDC)	Hydrographic surveying (11)

It needs to be emphasised that fields specified by Bannister et al. (1998) have matching fields in the *UDC*. Thus, the class of *Topographical surveys* (A) corresponds to *Topographic surveying* in the *UDC* (528.42), *Engineering surveys* (B) *to Engineering survey. Special fields of surveying* in the *UDC* (528.48), *Cadastral surveys* (C) to *Cadastral survey* in the *UDC* (528.44) and *Geographic information systems* (D) to *Cartography. Mapping* in the *UDC* (528.9).

1.2.3 Development of the classification system for surveying

The best way to develop a complete classification for the field of surveying is to compare the classification systems of the *UDC*, *EB* and the classification derived by Bannister et al. (1998), as further refined and presented in Table 1-1. The classifications of surveying fields of the *UDC*, *EB* and the refined classification by Bannister et al. (1998) are presented in Table 1-2. For the *UDC* and *EB,* apart from the names of surveying fields, classes under which they occur have been provided.

Table 1-2 Juxtaposition of surveying fields in the UDC, Britannica and in Bannister et al. (1998)

Fields of surveying				
UDC	**Code**	**EB**	**Code**	**Bannister et al. (1998)**
Theory of errors and adjustment	528.1			Analysis and adjustment of errors
Figure of the Earth. Earth Measurements. Mathematical, physical and astro-nomical geodesy	528.2	Astronomical observations. Navigational techniques and devices	723. G.6., 723. G.7.	Satellite positioning systems
Geodetic surveying	528.3	Surveying	723.G.1	Geodetic surveying
Topography	528.42			Topographic surveys
Cadastral survey	528.44			Cadastral surveys
Hydrographic and coast surveying	528.47	Oceanographi c surveys (special field survey)	723.G.4.	Hydrographic surveying
Engineering survey	528.48			Engineering surveys
Geodetic instruments and equipment	528.5			
Photogrammetry. Methods and instruments	528.7			Photogrammetry
Cartography. Mapping	528.9	Cartography and mapping	723.G.2 723.G.3	
				Geographic information systems

Table 1-2 shows the differences between the three classification systems. The *UDC* contains ten out of eleven fields specified in the above table, the *EB* includes only four fields and the surveying classification developed in Table 1-1 has nine out of eleven fields. The reason why the *EB* contains the smallest number of surveying fields is not that it is less complete than the other classification systems but because it is less explicit. Surveying information in the *EB* is hidden in various sections, e.g. *Analysis and adjustment of errors* is included in Part Ten, Division *Mathematics,* Section *Application of Mathematics*. Table 1-3 presents surveying subfields from Table 1-1 contrasted with the corresponding sections in the *UDC* and *EB*. It is a starting point for devising a complete classification system for the subfields of surveying.

While Table 1-2 was based primarily on the information included in Table 1-1 and its aim was to find corresponding fields in the *UDC* and *EB*, Table 1-3 provides a complete overview of the subfields of surveying in these three sources. Therefore, there is a separate row in Table 1-3 for *Geodetic instruments and equipment* and *Cartography*.

Most of the subfields occur in all classification systems, although they may be a part of broader fields and therefore have different names. For example, *Analysis and adjustment of errors* from Bannister et al. (1998) is *Theory of errors and adjustment* in the *UDC* and *Application of Mathematics* in *Britannica*. Similarly, *Satellite positioning system* is part of the class *Figure of the Earth. Earth measurements. Mathematical geodesy. Astronomical geodesy* in the *UDC* and corresponds to *Astronomical observations. Navigational techniques and devices* in the *EB*.

The subfields *Geodetic surveying*, *Topographic surveys*, *Cadastral surveys* and *Engineering surveys* have corresponding fields in the *UDC* that have nearly identical names. However, *Britannica* is not so specific about the types of surveys and has just one general category *Surveying* which includes all types of surveys.

Table 1-3 Missing surveying fields and codes in the *EB*

Fields of surveying				
UDC	Code	EB	Code	Bannister et al. (1998)
Theory of errors and adjustment	528.1	Application of Mathematics	10/23.	Analysis and adjustment of errors
Figure of the Earth. Earth Measurements. Mathematical geodesy. Physical geodesy. Astronomical geodesy	528.2	Astronomical observations. Navigational techniques and devices	723.G.6 723.G.7	Satellite positioning systems
Geodetic surveying	528.3	Surveying	723.G.1	Geodetic surveying
Topography	528.42	Surveying	723.G.1	Topographic surveys
Cadastral survey	528.44	Surveying	723.G.1	Cadastral surveys
Hydrographic and coast surveying	528.47	Oceanogra-phic surveys (special field survey)	723.G.4	Hydrographic surveying
Engineering survey	528.48	Surveying	723.G.1	Engineering surveys
Geodetic instruments and equipment	528.5	Instruments for measuring basic dimensions, instruments for measuring physical properties and relationships derived from basic dimensions	723.D.1723.D.2	(-) Geodetic surveying
Photogramme-try. Methods and instruments	528.7	Mapping and cartography (aerial surveys)	723.G.2.	Photogramme-try

Cartography. Mapping	528.9	Cartography and mapping	723.G.2.	Geographic information systems
Cartography. Mapping	528.9	Cartography and mapping	723.G.2723.G.3	(+) Cartography

The class of Hydrographic surveying has a similar category of Hydrographic and coast surveys in the UDC and a more specific category of Oceanographic surveys in Britannica. The UDC specifies Geodetic instruments and equipment as a separate category and the EB lists Instruments for measuring basic dimensions, instruments for measuring physical properties and relationships derived from basic dimensions. This category is not specified in Table 1-1 as a separate category but is a part of the Geodetic surveying class and for this reason it will not form a separate category in the classification of the subfields of surveying (it is marked in bold font to signal that it does not occur in Table 1-1 and has a minus sign, which indicates that it has to be removed from the new classification system).

The subfield of Photogrammetry is represented as Photogrammetry. Methods and instruments in the UDC classification and is a part of the category Cartography and mapping (aerial surveys) in the EB.

The same field *Cartography and mapping* in *Britannica* has a corresponding field *Cartography. Mapping* in the *UDC*. The range of fields it covers is quite broad, because it deals with everything that relates to maps. Although *Photogrammetry is listed* as the first of its subfields (keeping the numerical order of the *UDC* classification), the basic reference has to be made to *Cartography*, which is a separate subfield of surveying that has not been specified in Table 1-1 but needs to be added to a new classification (it is marked in bold font to signal that it does not occur in Table 1-1 and has a plus sign, which indicates that it has to be added to the new classification system). The field *Cartography and mapping* also incorporates the class of *Geographic information systems* from Table 1-1.

It is important to note that different classification systems use designations *survey* and *surveying* interchangeably in the names of surveying fields. For the purpose of consistency, I will use *surveying* throughout the thesis.

A final classification of surveying developed on the basis of the classification by Bannister et al. (1998), the *UDC* classification and *EB* classification contains the following items:

1. Analysis and adjustment of errors
2. Satellite positioning system
3. Geodetic surveying
4. Topographic surveying
5. Cadastral surveying
6. Hydrographic surveying
7. Engineering surveying
8. Photogrammetry
9. Geographic information systems
10. Cartography

1.3 Selection of data categories

The aim of this section is to discuss the structure of my term record, which is a basic unit of the surveying termbases in which surveying terms are stored. This termbase provides the basis for the morphological and semantic analysis carried out in chapters three and four.

The main focus of this section is the selection and specification of data categories to be included in the surveying termbases. In order to decide on the type of information included in each term record in a surveying termbase, I look at existing term records in the literature on terminology and termbases for other fields (1.3.1), as well as at the features of data categories (1.3.2). Then, I decide which data categories to include in a surveying termbase and provide a detailed description of their features (1.3.3).

1.3.1 Overview of data categories

Maria Teresa Cabré (1999), Juan C. Sager (1990) and Sue Ellen Wright (2001) use different approaches to term records. They all refer to the terminological record, which is a complete entry in a termbase, while Cabré (1999, p. 123) additionally lists an extraction record and correspondence record. An extraction record is an outcome of the term extraction process. The correspondence record is specific to Cabré's approach. It is a type of record which correlates all the designations for a single concept as it links two or more monolingual databases in a correspondence database.

According to Cabré (1999, p. 124), standard, monolingual terminological records normally contain the following data categories:

- identification of the term (number of the record);
- entry term;
- grammatical category;
- subject area(s);
- definition;
- source of the definition;
- context(s);
- source of the context;
- cross-reference to synonymous terms, which may be highlighted in a definition or context example(s) and then repeated in a separate field, or may not occur anywhere in the corpus but will be given in a separate field in the term record;
- illustration;
- author of record and date when the record was written;
- miscellaneous notes for unanticipated information.

Cabré (1999, p. 125) claims that a terminological record may additionally include information on the standardisation body. This information is crucial in case of synonymy as the choice of one of the forms as the actual term and the other one as a deprecated term has to be confirmed in standards, e.g. ISO standards or British Standards.

Cabré (1999, p. 125) develops a sample terminological record for Catalan which, after minor modifications, can be used for surveying terms in English and Polish. The terminological record for the term *circular level* created on the basis of Cabré's template for terminological records is presented in Table 1-4. The surveying corpus compiled for the purpose of creating a surveying termbase was narrowed to three subfields: geodetic surveying, cartography and GPS (see chapter two for a detailed description) in such a way that separate subcorpora were created for these particular domains, which is reflected in the *source of the term* specification in Table 1-4.

Table 1-4 Sample monolingual terminological record according to Cabré's guidelines

entry term: **circular level**			record no: **1**
source of the term: **geodetic corpus (Bannister et al. 1998: 55)**	gram. category: **noun phrase (NP)**	standardisation body: **ISO 9849:2000**	
subject area: **geodetic surveying**			
def: **a level having the inside surface of its upper part ground to spherical shape** source: **ISO 9849:2000** synonym: **bull's eye level, box bubble, circular bubble** context 1: **If a circular level bubble on a total station does not remain centred when the instrument is rotated in azimuth, the bubble is out of adjustment.** source of context 1: **Ghilani and Wolf (2008, p. 215)** context2: **The tribrach consists of a minimum of three components, which are a clamping mechanism, levelling screws, and a circular level bubble.** source of context 2: **Ghilani and Wolf (2008, p. 214)**			

author of the record: **Ewelina Kwiatek**			
notes: ***Circular level* is a holonym of *level* (levelling instrument) and a hyponym of *level* (levelling instrument component).**			
date of the record: **10.07.2009**			

Cabré (1999, p. 124) makes a distinction between monolingual termbases with translation equivalents and bilingual termbases, which consist of monolingual records with correspondence records.

Sager (1990, p. 143) proposes a different structure for the term record. Information to be included in a monolingual term record is put under the following data categories:

- the source information which links the term record to the raw data files, from where the term, definition and context, as well as the associated information, have been extracted;

- the entry term, which is either a linguistic item, or a label for a concept, or both;
- the semantic and conceptual specification of the term which contains the definition, a subject field attribution, scope notes and links to other concepts;
- the linguistic specification of the term, which can give only variants and abbreviations (if minimal) or all manner of morphological and syntactic specification (if more complete);
- the pragmatic specification of the term, which covers examples of the context and usage notes;
- the housekeeping or administrative information such as the record number, the name of the terminologist and the dates of first processing and subsequent updating of the record.

Those termbases which are translation-oriented also include the foreign language equivalent (Sager, 1990, p. 144). Sager suggests a model for a terminological record which which is presented in Table 1-5 for the term *circular level.*

When comparing term records in Table 1-4 and Table 1-5, one may notice that they differ in the presentation of information and the level of detail. Source information in Table 1-4 is limited to the source of the term, while in Table 1-5 a whole section is devoted to source information, with separate fields designed for origin and type. Sager's record structure subdivides the information included in the record into conceptual, linguistic and pragmatic and has a separate section for housekeeping information (author, date, record number and a pool number, which identifies the subset of a database, e.g. the terminology of a particular product). Although Sager's record seems to be more specific than Cabré's record, it is not. It includes more source and housekeeping information and the usage note which gives information on the usage of the entry in context. However, its presentation of source information is not transparent when different sources have to be provided for different types of information. For example, in *linguistic specification* there are different sources for the term and synonyms. They have to be put together as only one field is provided which affects the presentation.

Table 1-5 Sample monolingual terminological record according to Sager's guidelines

SOURCE INFORMATION					
origin: **ISO 9849:2000**	type: **ISO stan-dard**	origin: **Bannister et al. (1998: 55) for the term; ISO 9849:2000 for synonyms**	type: **surveying textbook ISO standard**	origin: **Ghilani and Wolf (2008, p. 215) for example 1;** **Ghilani and Wolf (2008, p. 214) for example 2.**	type: **text-books**
No.	page	No.	page	No.	page
CONCEPTUAL SPECIFICATION		LINGUISTIC SPECIFICATION		PRAGMATIC SPECIFICATION	
		language: **English**		language: **English**	
definition: **a level having the inside surface of its upper part ground to spherical shape**		term: **circular level**		context: **Geodetic corpus**	
field: **Geodetic surveying**		grammatical information: **noun phrase**			
links to other concepts: **level**		synonym: **bull's eye level, box bubble, circular bubble**		usage note: standardised term	
scope notes: ***Circular level* is a holonym of *level* (levelling instrument) and a hyponym of *level* (levelling instrument component).**		abbreviation:		example 1: **If a circular level bubble on a total station does not remain centered when the instrument is rotated in azimuth, the bubble is out of adjustment.**	

subject field: **Geodetic surveying**		variants		example 2: **The tribrach consists of a minimum of three components, which are a clamping mechanism, levelling screws, and a circular level bubble.**	
date: **01.10.2008**	type: extracting terms	date: **01.10.2008**	type: providing definition	date: **02.10.2008**	type: finding examples in the corpus
record number: **1**		pool number		terminologist: **Ewelina Kwiatek**	
HOUSEKEEPING INFORMATION					

Wright (2001a, p. 553) refers to *ISO 12620: 1999 Computer applications in terminology – Data categories* (ISO, 1999) when discussing data categories to be included in the database. ISO Technical Committee 37, Terminology, Sub-committee 3 for Computer Applications, Working Group 1 has elaborated the above-mentioned standard by collecting a broad selection of the terminological data categories used in many different termbases.

ISO 12620 is divided into three major groups that reflect the major types of information contained in a termbase:

1. Terms and term-related data

Subgroup 1 consists of the data category *term* and contains a term or other information, such as a phraseological unit or standard text, treated as if it were a term.

Subgroup 2 specifies data categories for term-related information such as:

- indication that a term is a symbol, formula, equation, or a materials management category;
- indication that a term is a full form or some type of abbreviation, e.g. *sonar* is abbreviation of *sound navigation and ranging;*
- indication that a terminological unit is a certain type of phraseological unit, e.g. *field of view* is a noun phrase;
- grammatical information on the term;
- regional, temporal, register-related and proprietary restrictions;
- etymology, pronunciation, syllabification, hyphenation, and morphological information;

- specific term-related administrative information associated with the processing of terms in standards and language-planning environments.

Subgroup 3 specifies data categories for information relating to equivalence between or among terms assigned to the same or very similar concepts, e.g. *byway open to all traffic* is equivalent to *szlak turystyczny* 'tourist trail' but only in the context of tourism.

The most useful category in which equivalence can be included is the transfer comment, which is a free-form note in which users can record salient information on target language usage, directionality, or specific reliability-related concerns (Wright, 2001a, p. 563).

2. Descriptive data categories

Subgroup 4 specifies data categories for the classification of concepts into subject fields and subfields, along with other classification-related information.

Subgroup 5 specifies data categories for concept-related description, i.e. different kinds of definitions, explanations and contextual material provided to define or determine the subject field and concept to which a term is assigned.

Subgroup 6 specifies data categories for indicating relations between pairs of concepts (e.g. generic, partitive, sequential, temporal, spatial and associative).

Subgroup 7 specifies data categories used to express the position of concepts within concept systems.

Subgroup 8 specifies the data category note. This category stands alone because it cannot be associated with any of the other data categories.

3. Administrative data categories

Subgroup 9 specifies data categories for documentary languages and thesauri.

Subgroup 10 specifies data categories for other strictly administrative information.

Supplemental ISO 12200 Data Category Groups

Subgroup 11 specifies special codes used in the MARTIF (Machine Readable Terminology Interchange Format) standard.

Subgroup 12 specifies bibliographic data categories.

While Cabré and Sager provide a ready-made template with data categories, Wright simply lists types of data categories within the three groups: term and term-related data, descriptive data categories and administrative data categories. It is possible to develop the template for Wright's categories by analogy to Cabré's and Sager's templates. Thus, the template developed for *circular level* is illustrated in Table 1-6.

Table 1-6 Monolingual terminological record created on the basis of Wright's theory

TERM AND TERM-RELATED INFORMATION
term: **circular level** grammatical information: **noun phrase, singular** pronunciation: /ˈsər-kyə-lər le-vəl/ date of the first documented use: **1721** synonym: **bull's eye level, box bubble, circular bubble**
DESCRIPTIVE INFORMATION
subject field: **Geodetic surveying** definition: **a level having the inside surface of its upper part ground to spherical shape** context 1: **If a circular level bubble on a total station does not remain centered when the instrument is rotated in azimuth, the bubble is out of adjustment.** context 2: **The tribrach consists of a minimum of three components, which are a clamping mechanism, levelling screws, and a circular level bubble.** superordinate concept: **level** subordinate concept: coordinate concept: **tubular level** note: ***Circular level* is a holonym of *level* (levelling instrument) and a hyponym of *level* (levelling instrument component).**
ADMINISTRATIVE INFORMATION
author of the record: **Ewelina Kwiatek** date: **10.07.2009** cross-references: **level, tubular level** source_definition: **ISO 9849:2000** source_context 1: **Ghilani and Wolf (2008, p. 215)** source_context 2: **Ghilani and Wolf (2008, p. 214)**

When the term record in Table 1-6 is examined, it may be noticed that some data categories are common to all three term records presented above. These data categories include: the entry, subject field, definition, grammatical information, context(s), sources, notes, cross-references, synonyms, author and date. The content and the structure of the term records proposed by Cabré and Sager are similar. Cabré additionally includes such information as standardisation body or status code which may be given for all term-related information or only for some of them.

This data category is particularly useful for synonyms as it is important to refer to a standardised body to justify why one of the forms is considered the main form and the other one is a synonym. The term record by Wright contains a lot of information which was not included in the other term records being discussed here. The data, such as superordinate, subordinate and coordinate concepts, may be very useful in establishing links between concepts in a termbase. Therefore, I decided to include hyperonyms and holonyms in the surveying termbase. On the other hand, some information provided in the term record by Wright may be redundant in the surveying termbase. This information includes pronunciation and etymological information.

I have created a term record for surveying terms by selecting those data categories that occur in at least two of three approaches: record number, term, abbreviation, grammatical information, subject field, definition, context(s), sources, notes, cross-references, synonyms, author and date. Next, I assigned them to four groups: term and term-related information, descriptive information, concept-related information and administrative information. I enhanced this classification by adding data categories which I find particularly useful, e.g. standardisation body (status) from Cabré's record and hyperonym and holonym from Wright's record. The outcome of my work is presented in Table 1-7.

The term record in Table 1-7 contains an additional data category *entity type*, which has not been mentioned before. *Entity type* describes the ontological categories, like event, state, thing, time, amount, etc. to which all terms belong and which help in formulating definitions (Jackendoff, 1983, p. 51). The entity type is useful in defining terms for which hyperonyms and holonyms cannot be specified. There are several definitions starting from the entity type in the surveying termbases. Entity types: process, organisation, method, science and instrument are particularly frequent. e.g. *calibration* is defined as 'the process of bringing the optical elements of an optical system into proper relationship with each other' and *European Space Agency* is characterised as 'an intergovernmental organisation dedicated to the exploration of space, currently with 17 member states'.

Table 1-7 The term record created specifically for the requirements of surveying terminology on the basis of existing term records

TERM AND TERM-RELATED INFORMATION
record no: **1** term: **circular level** abbreviation: grammatical information: **noun phrase**
DESCRIPTIVE INFORMATION
definition: **a level having the inside surface of its upper part ground to spherical shape** subject field: **Geodetic surveying** source: **ISO 9849:2000** example 1: **If a circular level bubble on a total station does not remain centered when the instrument is rotated in azimuth, the bubble is out of adjustment.** source: **Ghilani and Wolf (2008, p. 215)** example 2: **The tribrach consists of a minimum of three components, which are a clamping mechanism, levelling screws, and a circular level bubble.** source: **Ghilani and Wolf (2008, p. 214)**
CONCEPT-RELATED INFORMATION
synonym: **bull's eye level, box bubble, circular bubble** status: **the term is used in ISO 9849:2000 with preference to 'bull's eye level, box bubble, circular bubble'** hyperonym: **level** holonym: **level (levelling instrument)** entity type:
ADMINISTRATIVE INFORMATION
notes: ***Circular level* is a holonym of *level* (levelling instrument) and a hyponym of *level* (levelling instrument component).** author: **Ewelina Kwiatek** date: **10.07.2009**

1.3.2 Issues in selecting and organizing data categories

Data categories have many features that affect the accuracy and effectiveness of data maintenance and manipulation. These features need to be specified prior to the termbase creation in order to safeguard the integrity of the data within the termbase. According to Schmitz (2008c), the following features should be documented before the process of creating a termbase begins:

- elementarity of data categories;
- granularity of data categories;

- redundancy of data within data categories;
- dependencies of data categories;
- modelling variance;
- choice of level for data categories.

The elementarity of data categories depends on the fact that only one kind of data and only one instance of that particular data category can be included in a given field (Wright, 2001a, p. 559). It means that the full form of the term, e.g. *electromagnetic distance measurement* and its abbreviation *EDM* are given in separate fields and only one abbreviation is given in the abbreviation field, even if a term has more than one abbreviation. Other abbreviations can be handled in additional abbreviation fields that need to be introduced if such a situation occurs. Breaking the elementarity rule would occur if data was presented as in (2).

(2) term: electromagnetic distance measurement (EDM).

The granularity of data categories refers to the difference between methods for subdividing information (Wright, 2001a, p. 557). The term *granularity* is an analogy drawn from the notion of grain size used in such fields as optics, where the higher the granularity, the smaller the grain size and the greater the precision. Low granularity in the case of the termbase would mean that all grammatical information including gender and part of speech is put in one data element. The granularity in the surveying termbase is higher, as gender and part of speech are presented in different data elements. Increased granularity facilitates retrieval and manipulation of data. Table 1-8 presents a term record for *nüvi® 1200 pocket street GPS* with minimum granularity, while Table 1-9 presents the same term record with increased granularity. The entry in Table 1-9 can be sorted by Instrument name, Instrument type, Manufacturer's name, Street address, City, State, etc., while the entry in Table 1-8 has much more restricted sorting options. Lower granularity in case of Table 1-8 limits data usability.

Table 1-8 Term record for nüvi® 1200 pocket street GPS with minimum granularity

Data element	Content
Instrument	nüvi® 1200 pocket street GPS
Manufacturer	Garmin International, Inc.
	1200 E. 151st Street Olathe, KS 66062-3426 USA

Table 1-9 Term record for nüvi® 1200 pocket street GPS with increased granularity

Data element	Content
Instrument	nüvi® 1200 pocket street GPS
Instrument type	pocket street GPS
Manufacturer's name	Garmin International, Inc.
Street address	1200 E. 151st Street
City	Olathe
State	Kansas
Zip code	66062-3426
Country	USA

Redundancy of data within data categories is the next problem that needs to be dealt with. Wright (2001a, p. 559) claims that a cardinal rule of database management is to avoid redundancy. Complete bibliographic references could be provided for each text element in the terminological entry (the term, definition, examples, and synonym). However, not all of this information is necessary. The terminologist has to bear in mind that if he/she provides references for most of the data elements and makes mistakes in some of them, he/she will have to spend a lot of time correcting them. It will mean finding each citation to the source and making the desired change. Moreover, providing all this information occupies storage capacity. It also requires additional maintenance and often slows down retrievability of the termbase. The problem of redundancy of data has been solved in the surveying termbase by providing short references (e.g. author, date and page) to only such elements as definition and examples. A separate database with complete bibliographic entries has been created in EndNote, which may be consulted if full references are needed.

Dependency of data categories should now be considered. Schmitz (2008c) claims that data categories are dependent on one another. Grammar is dependent on term, source is dependent on definition and the source of definition has to be differentiated from the source of the example. Dependency affects the selection and presentation of data categories in my surveying termbases. For example, English terms are followed by abbreviations and grammatical information, whereas for Polish terms, apart from grammatical information on part of speech, additional gender specification is provided for nominal entries. As for sources, the termbases have four separate data categories for sources; one data category for the source of definition and three data categories for sources of three examples.

Modelling variance is the next feature to be discussed. Wright (2001b, p. 581) identifies two fundamental varieties of data modelling. The first one refers to the situation when there is a distinction between different word forms so that one of them is recognised as the term, while others are perceived as synonyms or abbreviations. This situation is illustrated in (3), where *digital terrain model* is the main term, *DTM* is its abbreviation and *digital elevation model* is a synonym. Different word forms constitute separate data categories in the termbase.

(3) term: **digital terrain model**
abbreviation: **DTM**
synonym: **digital elevation model**

The application of the model presented in (3) means that the termbase has to include three separate data categories to keep information on main terms, abbreviations and synonyms.

The second variety of data modelling presented in (4) occurs when all word forms are recognised as terms and there is a further specification about the term type, which indicates whether the term is a main term, an abbreviation or a synonym.

(4) term: **digital terrain model**
term type: main term
term: **DTM**
term type: abbreviation
term: **digital elevation model**
term type: **synonym**

The solution outlined in (4) would require inserting *digital terrain model*, *DTM* and *digital elevation model* as separate entries in the termbase and adding a data category term type, which can be a picklist including three types of terms that can be selected for a particular entry.

The solution offered in (3) is more transparent as each term with its alternative word forms is treated as a separate entry in the termbase, while the solution in (4) multiples the number of entries which in fact refer to one term. On the other hand, (4) is more flexible as it provides multiple synonyms and abbreviations of synonyms.

Data modelling variance can refer to other data categories, for example gender. There are two modelling variants available. The first one, which is presented in (5), depends on creating one data category for gender with values (masculine, feminine, neuter or m, f, n) that can be selected from a picklist or which may be specified manually for each entry.

(5) gender: **m. / f. / n**

The second modelling alternative presented in (6) involves creating separate data categories for masculine, feminine and neuter genders with possible values yes/no which can be ideally selected from a picklist or assigned manually to each entry.

(6) masculine: **yes/no**
feminine: **yes/no**
neuter: **yes/no**

The variant presented in (5) is more economical as it requires one data category with three possible values, while the second variant depends on creating three separate data categories, one per gender, with two values. On the other hand, (6) offers additional possibilities such as multiple or no gender.

In the surveying termbase simple modelling solutions have been applied. Data categories related to term (i.e. citation form, abbreviation and synonym) function as separate data fields, as in (3). The modelling solutions for data categories *part of speech* and *gender* follow the pattern applied in (5).

Schmitz (2008c) also recognises choice of level as a feature of data categories. The typologies used in (3) and (4) indicate the level in the data model on which the data categories should be attached (concept-, language- or term-oriented data categories). In some cases the level is exactly defined, e.g. part of speech at the term level. In many cases, however, the choice of level depends on the objective and philosophy of the termbase, e.g. graphics can be used as resources for the whole entry, as a reference for a single language, or even to illustrate a single term. In the surveying termbase, the choice of level is reinforced by the terminology management system. All data categories are at the same level.

1.3.3 Description of data categories

Data categories include different types of information. All this information should be presented according to a set of standardised conventions in order to ensure efficient retrieval and exchange of the information (Cabré, 1999, p. 139). A data category name may suggest that a particular category contains a certain type of information but it is necessary to define or describe what type of information it is, so that users will understand what the content of the category really is (Wright, 2001a, p. 555). Descriptions of data categories provide explanations of these data categories which may be new or unknown to the user. For example, a non-expert may not know what the *status* of the synonym is, while an expert user may be interested what the possible values for status are. Thus, documentation of data categories has three important functions:

- it helps to explain basic notions used in the terminological work to non-expert users;
- it explains details of policy to expert users;
- it lays down the terminologist's policy to themselves.

The well-prepared documentation that is followed in terminological work prevents inconsistencies in the termbase. For example, it provides a set of rules for formulating definitions or criteria for the selection of examples.

According to Wright (2001a, p. 556), apart from specifying the content of the data entered into a data field, it is also important to determine the data type, i.e. text, dates, numbers, etc. Therefore, documentation should also include the specification of values that are available for data categories. Some of these values may be checked in standards. For instance, ISO 12620 (ISO, 1999) recommends the use of standard ISO date format (year-month-date, e.g. 2009-07-10) for dates. However, it is not a requirement, as many existing termbases use different formats. While designating the data type, one has to bear in mind standardised formats for representing information for country, language and even bibliographic references. For example, international language abbreviations are required and when a bibliographic reference is used in the termbase, it should be a short reference (e.g. author, data and a page number). Some databases may also specify the field size (the maximum number of characters that can be used within the field) for certain elements. The trend in terminology is not to restrict the length of such elements as term, definition and context.

Wright (2001a, p. 556) claims that delineating the data type and field length results in establishing the permissible instances, when only data of specific type and of limited length may be entered into the data fields. A picklist is a good example of permissible instances as it delimits the choice of values to a few items which have been predefined and their format has been determined (e.g. the gender picklist for Polish includes three values: *męski* 'masculine', *żeński* 'feminine' and *nijaki* 'neuter', where full names of gender forms have been included in the list). Permissible instances are very useful in electronic termbases as they save time and prevent inconsistencies in the data format.

Below a description of data categories in the surveying termbases is provided. It is accompanied by a discussion of possible value types for each attribute. It is important to note that two monolingual termbases, one for English surveying terms and one for Polish surveying terms have been created. These termbases contain a separate data category: translation equivalent, e.g. the English termbase provides Polish equivalents of English terms.

ID (record number)

A record number consists of the number of the entry, possibly with some subcategories indicating the general origin of the whole record (Sager, 1990, p. 153). In the surveying termbase ID is simply a reference number which is entered automatically by the terminology management system when a new entry is created. ID is an ideal solution for computers as it is created automatically and gives users control over the information, e.g. sorting smallest to largest and largest to smallest, searching facilities. In MS Access, which I use for the termbase, ID is an Auto Number whose value is a long integer starting at one and incremented by one. Auto Number is a useful solution as it is unique which means that MS Access never gives the same value to two records. It cannot be used as an indicator of the number of records in the database, however. If an existing record is deleted from the database, Access never reuses that Auto Number value again. As a result there may be more Auto Number values than actual records in the database.

Term

Terms are identified by entry forms, called *entry terms*. According to Sager (1990, p. 143) the field *term* should be filled with citation forms, as conventionally used in dictionaries (e.g. singular for nouns, masculine for gender-inflected adjectives, active infinitive for verbs, lower case initials for most of the terms apart from organisation names and abbreviations that are more common than the full forms of the terms).

The abundance of word forms that represent the same concept is another problem concerning terms. If there are a few terms with the same definition, it is often difficult to choose one term as the entry term and declare others as synonyms. This problem is solved by the data category *status* containing the name of the standard body that recognises one form as the main form and the other one as the synonym. The solution is documented further down in this section.

There are certain conventions for entering terms which, if followed, preserve the consistency and accuracy of data in a termbase. The lists of such conventions for English terms were elaborated by Wright (1997, p. 16-17) and Schmitz (2008c). They include the following items:

- terms should normally be lower case;
- only those terms that are capitalized in standard discourse are capitalized in terminological databases, i.e. proper nouns and proper adjectives are capitalized (e.g. *Global Positioning System*, *Very Long Baseline Interferometry);*
- nouns should be entered in their singular form. In the case of terms with irregular plural forms, these forms may be given in notes. For example, in the surveying termbase there is a term *ephemeris*, whose irregular plural form is *ephemerides*;
- nouns that only occur in the plural or that have different meanings in the plural are an exception to this rule. For instance, the English singular noun *plastic* means something different from the plural form *plastics* (types of plastic materials). In such cases, the singular and plural forms should be treated as separate concepts;
- verbs should be entered in their infinitive form without the particle *to* before the verb;
- multiword terms should be entered in their canonical form, e.g. *apparent solar time*, *geostationary orbit.*

Wright (1997, p. 22) also states that terminologists should always check standard practice for the languages of the termbase as conventions in different languages may vary. Most of the conventions listed above are quite universal but there is a list of principles for entering the term that applies to Polish only and does not concern English. It includes the following items:

- if there is no standardised Polish equivalent for the English term, the English term may be used as the entry term and a Polish translation or a description of a concept should be provided in the notes, e.g. there are no Polish equivalents for such terms as Selective Availability and Anti-Spoofing;
- in Polish, adjectives referring to nationalities are not capitalized if they are not part of proper names, e.g. narodowość polska 'Polish nationality', amerykański odbiornik GPS 'American GPS receiver';
- Polish nouns have gender, which is specified by the ending of the word. Feminine nouns end in a, neuter nouns end in o or e, and the masculine gender has a consonant ending. There are exceptions to this rule, however. For example, satelita 'satellite', which has a feminine ending, is masculine. Due to the occurrence of exceptions, gender has to be specified for each noun;
- verbs are entered in their infinitive form and it is easy to distinguish them from other parts of speech as they end with 'ć', e.g. mierzyć 'to measure', niwelować 'to level'. Polish is a highly inflective language. The Polish infinitive comes from the base word, e.g. mierz and the verbal affix -yć. The affix changes in the conjugation and the base word stays the same (mierz-ę 'I measure', mierz-ysz 'you measure', mierz-y 'he measures').

Grammatical information

Grammatical information covers conventional information contained in dictionaries (Sager, 1990). This information depends on the language of the termbase. For the English surveying termbase it covers only part of speech. Grammatical information is a picklist which restricts the user's choice to n (noun), v (verb), adj (adjective) and adv (adverb). All the terms in the surveying termbases in English and Polish are nouns and noun phrases, therefore I added a Noun Phrase as an additional category to a picklist to distinguish those items that are multi-word expressions with the noun playing the role of the head of the phrase. Any irregular plural forms of nouns are entered in the data category notes due to the fact that few nouns have irregular plural forms.

Polish nouns and noun phrases, which inflect in cases, are given in the Nominative. Information on gender, which occurs for all nouns and noun phrases, can be selected from picklists. The gender category in the Polish termbase specifies three genders: *męski* 'masculine', *żeński* 'feminine' and *nijaki* 'neuter'.

Subject field

Dubuc and Lauriston (1997, p. 81) claim that no term exists without reference to a particular domain and therefore each term should be assigned to a subject field. The subject field assignation depends on a previously established classification, which should be an authoritative classification appropriate for the structure of the field (Cabré, 1999, p. 140).

The classification of terms by subject areas is a useful facility as it helps to deal effectively with large quantities of terms (Sager, 1990, p. 147) which can be, for example, sorted by this data category. The information on a subject field complements definition and is crucial in building a concept structure for a particular domain.

On the basis of various classification systems that can be used for the area of surveying, I managed to develop a classification system for surveying. The development of this system, which relies on the *UDC* classification, *The New Encyclopaedia Britannica* classification and the classification system extracted from the text book by Bannister et al. (1998) was described in detail in section 1.2.3. This classification system includes the following subject fields in the domain of surveying:

1. Analysis and adjustment of errors
2. Satellite positioning system
3. Geodetic surveying
4. Cadastral survey
5. Topographic surveys
6. Engineering survey
7. Hydrographic survey
8. Photogrammetry
9. Geographic information systems
10. Cartography

Subject field specification is a separate data category in the surveying termbase. It is a picklist containing the subject fields specified above.

Context

Exemplification of the usage of the entry term in a context is a good way of showing any peculiarities of a word form, inflection or collocation. The context should be chosen so that it complements the information provided in the definition. Dubuc and Lauriston (1997, p. 81) argue that the context plays a double role: it provides proof that the term is used in the particular field and it allows the terminologist to determine the relationship between a term and the subject field it belongs to. Cabré (1999, p. 138) distinguishes between different types of context for terms:

- testimonial contexts, whose purpose is to prove that a term occurs in a text. This type of context neither complements a definition nor locates the concept covered by the term within a concept system, e.g. *The mountains offer challenges for aerial photography*;
- defining contexts, which provide information about the meaning of the term. Contexts of this type may serve as a basis for formulating true definitions, e.g. *Spherical Error Probable (SEP) is the spherical equivalent of CEP, that is the radius of a sphere, centred at the actual position, that contains 50% of the three dimension position estimates;*
- metalinguistic contexts, which present the term as a unit within a formal system and create an approximate image of the concept covered by the term. Dubuc and Lauriston (1997, p. 82) call this type of context 'explicative contexts', e.g. *Cadastral data contains information about land parcel boundaries for all freehold and crown land parcels within Western Australia.*

Cabré (1999, p. 139) states that the best types of contexts in terminology are defining contexts, which is certainly true for definitions. Sager (1990, p. 150) is not so specific about the types of contexts and only claims that the context should complement the information provided in the definition and the usage note. The context given for general language entries is a mixed context, between a defining and a metalinguistic context (Cabré, 1999, p. 231). In the surveying termbases, I rely mainly on defining and metalinguistic contexts. I use defining contexts to formulate definitions and metalinguistic contexts to provide examples. In cases when finding any of these contexts is not possible, I use the testimonial context if it has the form of a sentence.

Dubuc and Lauriston (1997, p. 85) state that choosing the right context is a crucial part of terminological research. In corpus-based approach all examples should preferably be taken from the corpus. If the corpus does not have a sufficient number of examples, or does not have relevant examples, examples can be sourced from outside the corpus. There are three criteria that should be met when selecting a bibliography for terminological research (Dubuc & Lauriston, 1997, p. 85): representativeness of the text (the text should reflect usage of terms in the specialised field), the nature of the publication (the type of publication, i.e. whether it is a textbook, a monograph or a service manual) and a minimal level of presentability and reliability (only well-written texts are a relevant source of information for drafting valid term records).

The documents outside the corpus should reflect usage of the terms in specialised fields. They should contain a meaningful content, which sends the terminologists to such sources as monographs, textbooks, handbooks and service manuals. Examples in the termbase have to be followed by sources which indicate their origin.

According to Dubuc and Lauriston (1997, p. 87) and Wright (2001a, p. 564), there is disagreement among terminologists as to whether examples should be classified as concept-related or term-related information. The prevailing view advocated by many authors in the *Handbook of Terminology Management* including Dubuc and Lauriston is that the role of examples is to orient a concept and its designated terminological unit(s) to, and within, a specific subject field (Wright, 2001a, p. 564). Wright (2001a, p. 562) classifies examples as concept-related information which along with the subject field and definition belong to a wider group of descriptive information. The second view is that contexts provide term-related information in the form of collocational and other discourse-related features. Discourse-related aspects are useful only when the orientation of the context to the subject field has been fully established. The second view is adopted in general lexicography, concerned with the creation of general language dictionaries and in specialised lexicography, which is limited to collecting the terms of a particular domain for informative and descriptive purposes (Cabré, 1999, p. 38).

In the surveying termbase, three examples are provided for each term. The first example typically includes a term as it occurs in the corpus and should not be modified. The second and the third example can be modified, which means that, if they include pronouns or hyperonyms of the term, they may be used as examples, but the actual terms must be put in brackets next to the pronoun/hyperonym, e.g. *They (charts) are generally drawn on Gerard Mercator's projection (first introduced in 1569), which transforms the curved surface of the Earth onto a flat plane.*

Definition

Definition is a data category that has many different designations. Longworth (2006, p. 409) claims that a definition is a product of the activity of explaining to an audience the meaning of an expression. Following Aristotle, he argues that every definition includes two parts: the word that ought to be defined *(definiendum)* and the defining words *(definiens).* For example, in the sentence "Fortnight is a period of 14 days", *fortnight* is definiendum as it needs to be defined and *a period of 14 days is* definiens as it defines *fortnight.*

Hanks (2006, p. 399) indicated that, traditionally, definitions are applied to classes of entities which may be physical objects or abstract concepts. A traditional definition consists of a *genus* term, which specifies what sort of thing the entity is, and any number of *differentia,* which distinguish the entity from members of related sets. Thus, in the definition of *badgerdog,* the genus term will be *dog,* while the differentia will include information on its size (small), its colour (red), its hair (smooth) and its role (to scent, chase and flush out badgers). This tradition goes back to Porphyry (c. 232-303 A.D.), who wrote an introduction in Greek to Aristotelian logic. Definitions in Aristotle's view are more elaborate than in Hanks' specification as apart from genus and differentia, they also include species (specification of species to which an entity belongs), properties (features that distinguish a given species from other species within the genus) and accidents (other particular characteristics).

Another traditional delineation of definition was created by Plato, who claims that there is a group of objects which are knowable, unchanging idealisations and not subject to change over time. These objects may be defined by using necessary and sufficient conditions. For example, the ideal triangle is always a triangle. Necessary conditions for being a triangle include being a geometrical figure, having three and only three straight lines, and having three angles. What is more, these three conditions constitute a sufficient condition for being a triangle. Thus, if something that has three lines and three angles is encountered, it must be a triangle.

Although traditional approaches to definition are still valid and are applied by many contemporary linguists, some alternative ways of defining have also been proposed. For example, the *Longman dictionary of contemporary English* (Procter, 1978) uses a restricted defining vocabulary of only about 2,000 words to define its entries. Anna Wierzbicka, the Polish-Australian philosopher of language, proposes definitions based on 'universal semantic primitives' or 'atomic units of meaning' by which she understands words that are undefinable and which cannot be decomposed and explained in other words (Wierzbicka, 1996). The list of Wierzbicka's primitives includes 55 items that belong to 16 categories. For example, the category *Substantives* includes primes: *I*, *you*, *someone/person*, *people*; the category *Evaluators* includes primitives: *good* and *bad*. Wierzbicka's view is considered by many to be controversial, because although the meaning is defined in terms of primitives, the discussion is in ordinary English (Hanks, 2006, p. 401).

Many contemporary designations of definition not only outline features of definition but also stress its role in language. Cabré (1999, p. 104) refers to two ISO standards: ISO 704:2009 (ISO, 2009) and ISO 1087-1:2000 (ISO, 2000a) to explain what the definition is. ISO 704:2009 states that a definition is a complete description, usually in language, of a concept using other, already known concepts. ISO 1087-1:2000 designates a definition as a statement which describes a concept and which permits its differentiation from other concepts within a conceptual system.

Sager (1990, p. 146) claims that a definition is a bridge between the concept and the term. Many names of concepts, particularly those borrowed from other languages, are established by consulting definitions, e.g. *ambiguity resolution* is translated into Polish as *nieoznaczoność cykli fazowych* 'ambiguity of phase cycles', and is based on the definition of the concept as a direct translation of the English term does not make sense in Polish.

Classifications of definitions should now be discussed. Sager (1990, p. 45) enumerates many different types of definitions and finds three of them to be the most useful in terminology:

- analytic definition (genus at differentia), which relates a term to its superordinate, e.g. *circumference* 'a perimeter of a circle';
- synthetic definition, which identifies the place of the concept in a system of relations and mentions the subordinate terms, e.g. *cadastral map* 'a large-scale map established by cadastral survey showing the boundaries and dimension of the ownership of land';
- denotative definition, which lists all the subordinate terms, thus covering the extension of a term, e.g. *levelling instruments* are *compensator level, digital level, dumpy level, spirit level, tilting level, wye level*.

An analytic definition is very brief. It simply gives *genus*, which is a broader category to which the concept belongs (hyperonym), and *differentia*, which are the distinguishing characteristics of this concept. By contrast a synthetic definition is more descriptive, identifying all possible relations the concept has with other concepts. A denotative definition is completely different from analytic and synthetic definitions as it does not provide any description of the term but simply lists its instances.

Cabré (1999, p. 104) uses a different classification system for definitions. She differentiates three types of definitions:

- linguistic definitions;
- ontological definitions;
- terminological definitions.

These three types of definitions differ in the object they describe and the contents they express. The object of a linguistic definition is the linguistic sign. Linguistic definitions are aimed at distinguishing different concepts and therefore they include only those characteristics which are the most important in differentiating concepts. The object of an ontological definition is the class of objects in the real world. Ontological definitions cover all the particular intrinsic, extrinsic, essential and complementary aspects of a concept. An example of an ontological definition is an encyclopaedic definition. The object of a terminological definition is a concept of a special subject field. Terminological definitions describe concepts with reference to a special subject field. These three categories are not mutually exclusive as ontological definitions can be terminological or linguistic. The view that Cabré (1999) has of terminological definition is consistent with Sager's claim (1990), according to which definitions are usually complemented by subject field classification and apply to the specified subject field only.

Apart from these classification schemes for definitions that have been already discussed, Cabré (1999, p. 105) also specifies that definitions can be extensional and intensional. Intensional definitions bring together characteristics describing the concepts and are the same as synthetic or analytic definitions. Extensional definitions, which have been already listed by Sager (1990) as denotative definitions, enumerate the specific objects that a concept represents.

Both Sager (1990, p. 51) and Cabré (1999, p. 105) recognise the need for using terminological definitions in termbases. The definitions in the surveying termbase are also terminological.

Bowman et al. (1997, p. 215) as well as Cabré (1999, p. 104) and Schmitz (2008a) define a set of rules that must be followed to formulate terminological definitions. Definitions should be collected from reliable sources and can be quoted from standards when possible. The sources of the definitions should be provided. Definitions should be based on the superordinate concept for generic and partitive relations (Bowman, Michaud & Suonuuti, 1997, p. 215). In the surveying termbase, there are three types of definitions:

- definitions starting from hyperonyms;
- definitions starting from holonyms;
- definitions starting from entity type.

Hyperonymy is a semantic relation between a more general and a more specific word. e.g. *antenna* is a hyperonym of *choke-ring antenna.* An example of a definition starting from a hyperonym is *choke-ring antenna* 'the antenna consisting of several concentric rings around the actual antenna that create an electromagnetic field around the sides and the base of the receiver and effectively block multipath signals entering the receiver from below or at a low angle'.

Holonymy is a semantic relation between a word denoting the whole and a word denoting a part of the whole, e.g. *GPS* is a holonym of *ground segment*, *space segment* and *users*. An example of a definition starting from a holonym is *antenna* 'the part of the GPS receiver hardware which receives the incoming L-Band signal'.

Entity type describes the type of ontological category that the conceptual constituent represents such as: EVENT, STATE, THING, PROPERTY, PLACE, PATH, TIME, AMOUNT, etc. (Jackendoff, 1983, p. 50). An example of a definition starting from an entity type is *British Cartographic Society* 'an organisation that associates individuals and organisations dedicated to exploring and developing the world of maps'.

Bowman et al. (1997, p. 215) advise avoiding synonyms in definitions. Furthermore, if, for example, the term being defined is a noun, the definition of this term should start from a noun too. Highly technical expressions and mathematical formulas should be avoided as definitions must be clear and useful for the intended reader. Only one concept ought to be described per definition and only one good definition for each term should be provided.

Schmitz (2008a) specifies a set of criteria that the language used to write definitions should meet. He claims that the language used to write definitions should be formal; neither slang nor jargon are allowed. Such expressions as: *means*, *the term used for*, *the concept denoting,* are not allowed in definitions. Definitions should not include examples or expressions pointing to a prototype, e.g. *usually*, *as a rule*. The terminologist must also avoid circularity in definitions which means they should refrain from using the term being defined. Negative definitions should also be avoided as the requirement is to describe what the concept is, not what it is not. All these principles have been followed when writing up definitions for terms in the surveying termbase.

Sources

Sources are needed for the entry term, the definition, the contexts, translation equivalents and for synonyms (Sager, 1990, p. 151). The source of the entry should be as authoritative as possible. It can consist of the following types of information:

- source origin, which can be either a simple coded reference to a bibliographical database or a mnemonic code reference to a limited set of commonly used sources. It is worth noting that various types of sources will have different coded references, e.g. the book will be coded as author, year of publication and a page, while the standard will have the standard title, date of publication and the page if available;
- source type, which is the type of document the source comes from, e.g. ISO;
- source reference code or number, which is a reference to a separate source reference file in a large database. It gives full bibliographical details for written sources.

In the surveying termbase sources have been provided for definitions and all examples and synonyms. They have the form of short bibliographic references, while the full references are available in the EndNote database. Translation equivalents which are recognised as standard terms in the target language do not need sources. A lack of standard translation equivalents is often caused by conceptual mismatches (discussed in detail in 4.3), and finding a counterpart in the target language is usually a very difficult process based on the analysis of concept systems in the source and target languages that is carried out on the strength of information contained in numerous sources. Therefore, providing one source for the translation equivalent is not possible in such cases.

There is also a set of rules which have been defined by Schmitz (2008a) and which I followed in order to document sources properly. I used only quality sources such as textbooks, journals, technical magazines, manuals, and education websites. I give detailed and precise information on all sources consulted so that the reader is able to locate the information used and to check, if necessary, the evidence on which any discussion is based. For books, I provided the author, date, and a page number; for web sources, URL and the date when the source was accessed, or if possible, the author of the website, the name of the website, date when it was created or updated and the date when it was retrieved. Sources are given in separate columns that follow columns with definition and examples. A database of all references that occur in the surveying termbase has been created using EndNote and contains full bibliographic references.

Synonyms

Synonyms are numbered among informative cross-references whose purpose is to refer a given term to another term to broaden the information about its designation or concept or to show its relationships with other forms and concepts in the same field. In terminology the role of synonyms is to refer to other terms that designate the same concept (Cabré, 1999, p. 142). The surveying termbase contains a data category called *synonym*. There is also a data category called *status*, whose aim is to specify on what basis one of the forms is considered the main form (the term) and the other one a synonym.

Status

The status label is a code whose aim is to provide information on the quality of terms. For standardised terms, ISO recommends using the name of the standardising agency as the status label. The status label can apply to the whole record, to the entry term or to some information on the record (Cabré, 1999, p. 144). In the case of the surveying termbase it applies to the data category *synonym*. The status label gives such names as: ISO (International Organization for Standardisation), IEC (International Electrotechnical Commission), BSI (British Standards Institute) or Polski Komitet Normalizacyjny 'Polish Committee for Standardisation', i.e. Polish Standardisation Committee. The role of the status label is to indicate the standardised body which recognises one of the forms of the term as the main term and the other one as the synonym.

Abbreviation

Cabré (1999, p. 110) states that synonymy occurs between the following cross-references:

- a full form of a term and an abbreviation consisting of non-initial letters, e.g. *millimetre – mm*;
- the full form of a term and its initials, e.g. *electromagnetic distance measurement - EDM* or acronyms (initial syllables), e.g. *POLREF* (POLish REference Frame).

Abbreviations and initials (or acronyms) are prescriptive cross-references. Their aim is to indicate the existence of alternatives on the same level. They are treated together in the surveying termbase and belong to the data category *abbreviation* which is a separate category from synonyms.

Equivalent

The data category *equivalent* is an additional category introduced to the termbases. As the approach taken to compile termbases is concept-oriented rather than term-oriented, finding an equivalent is not the major task of the termbases but the outcome of the process of identifying the position of a concept within a concept system, which takes place by providing the definition of the term, subject field specification, hyperonyms and holonyms.

Most terms in the surveying domain have standardised equivalents but there are also quite a few terms that seem not to have any corresponding terms in the target language. Lack of equivalence may be a sign of a conceptual gap, which occurs when concept systems of two languages differ, and requires an in-depth analysis of the concept systems of the two languages to confirm whether this is the case.

Notes

Notes include additional information which cannot be attributed to any other fields. Notes can include a specification of subject or register, and their aim is to indicate a special field of application of the entry term. They can also give information about the usage of the entry term in context which cannot be provided in the form of examples of a real context. In the surveying termbase, they are broadly used for translation solutions offered in cases when there is no translation equivalent for a particular term.

Author

The record must contain coded identification of a person or group of people who have written it. In the surveying termbase it is the name of the terminologist. The name of the editor also occurs in sources if a major modification is made to a definition or examples.

Date

Date of record is the date of the production of the first record and any subsequent up-dates. In the surveying termbase it is the date when a record was completed. It is entered according to the format 2009-04-21, which is suggested in ISO standard and is available in the software package that was used for terminology management.

1.4 Software aspects of terminology management

Terminology management has undergone significant evolution from file cards and glossaries used until 1965, through mainframe terminological data banks such as Termium or Eurodicatum (1965-1975), mini-computer based terminological data bases, e.g. Ericsson, Danterm (1975-1985), PC-oriented simple terminology management systems for single users (1985-1995), sophisticated terminology management systems (TMS) for PC networks (1995-2000) up to web-based client-server terminology management systems (TMSs), such as MultiTerm which are currently used (Schmitz, 2008b). Thus, when speaking about terminology management nowadays, terminology management programs are referred to. They are defined as software products, designed for the management of terminological data, which enable the user to collect, store, manipulate, and retrieve terminology (Schmitz, 2001, p. 539).

This section is confined to the selection of the terminology management system which will suit the needs of creating the surveying termbase. Therefore I look at the criteria for evaluating terminology management programs (1.4.1) and analyse existing terminology management solutions according to these criteria (1.4.2). Then, I provide a description of the termbase I created using the selected software package (1.4.3).

1.4.1 Criteria for evaluating terminology management systems

Criteria for evaluating terminology management programs may be determined only if the purposes of applying the program, its outcome and envisaged users have already been specified. The aim of using software for terminology management is, in my case, to generate two separate termbases: one for English surveying terms and one for Polish surveying terms. These termbases are originally designed to be separate applications. They will be a starting point for morphological and semantic analyses of terms and concepts in chapters three and four. However, the ultimate goal of the project is to convert these termbases into a surveying dictionary. This dictionary aims to identify and solve conceptual mismatches, which are not uncommon in surveying terminology. The envisaged users of this dictionary are surveyors and translators.

Schmitz (2006, p. 586) and Wright (2001b, p. 579-583) claim that there are two basic principles which must be taken into account when defining a terminological entry structure in termbases: concept orientation and term autonomy. The key principle of concept orientation is that all terminological information belonging to one concept, including all terms in all languages and all term-related and administrative data, must be stored in one terminological entry. Term autonomy guarantees that all terms belonging to one concept should be managed (in one terminological entry) as autonomous (repeatable) blocks of data categories without any preference for a specific term. It can be done by designing a data model that allows the user to create an unlimited number of term blocks which contain individual terms and all additional data categories describing the term (e.g. abbreviation, synonym).

Apart from these two fundamental criteria, Schmitz (2001) mentions many other aspects that need to be considered when deciding on a terminology management system. In his view, the suitability of the software for the intended terminology task is the most important criterion. He distinguishes between terminology management systems with defined (fixed) structure and those with definable entry structure. In systems with fixed structures the data categories are pre-defined and the user cannot change the format of the terminological entry or the assigned length of the data fields. Database programs with fixed data entry structures are usually SQL-based relational databases. Terminology management systems with freely definable entry structures allow users to define their own data categories and their own data structure.

It is also important to check whether a terminology management system supports all the necessary terminological and administrative data categories, and whether the system can store the required volumes of data within data categories and handle multiple data of the same type.

The system's support for data integrity is another important technical aspect. Validation procedures are very useful as they check whether types of data and formats of data are allowed within data fields. If validation procedures are applied, the system warns the user about the incorrect format of the data. In some cases the system may correct the format of data to the format that has been predefined by the user when the termbase structure was designed.

Terminology management programs are often used together with word processors or translation memories so, depending on the type of the project, it may be useful to consider multitasking terminology management systems that can co-operate with other applications.

Another important category in the evaluation of terminology management programs covers user interface aspects such as screen displays, input and retrieval of information and information transfer (Schmitz, 2001, p. 545). The information displayed on screens should be clearly arranged and easy to read. A terminology management system should provide convenient input and retrieval procedures that facilitate efficient terminology work, e.g. picklists for closed data categories such as gender, or searches in data categories other than the term category. Since data management programs co-operate with translation memories or word processors, it should be relatively easy to transfer selected parts of the information from the termbase or the whole termbase to the format that is compatible with the selected application.

To sum up this discussion, there is a set of criteria such as concept orientation, term autonomy, entry structure (fixed entry structure vs freely definable entry structure), data categories, data integration, display of information, input and retrieval and transfer of information that are necessary to consider when searching for a terminology management system for this project. As my ultimate goal is to create a dictionary in which I present the differences between English and Polish conceptual systems, the criteria of the multitasking character of TMS is not so crucial. It could be important, however, if at some point I decided to use the termbase as the basis for an application that supports computer-assisted translation.

1.4.2 Terminology management solutions

Wright (2001b, p. 575) distinguishes three types of terminology management systems: word-processing programs, off-the-shelf database systems and dedicated terminology management systems. The usefulness of these systems for the purpose of creating the surveying termbase will be evaluated on the basis of criteria specified in the previous section.

Word-processing programs are word processors and spreadsheets such as Microsoft Word and Microsoft Excel. They allow the user to open multiple screen windows and work with a few texts at the same time. They provide searching, sorting and merging facilities that can be used to generate alphabetical lists and to manage text segments. Word-processing programs create tables with columns for different data categories. The structure of a termbase created in word processors or spreadsheets is flexible, as adding or deleting entries and data categories is very straightforward. The system supports all data categories. This solution, however, has a number of limitations. It is not suitable for termbases with many data categories, as the display of information on the screen may be very inconvenient if the table has many columns. The input and retrieval of information in such systems is very difficult as no automation is supported, e.g. by creating picklists. The system does not support data integration and validation. Concept orientation or term autonomy is not provided either. The user can try to define them, but these aspects may be difficult to maintain in the case of extensive termbases with many entries and many data categories.

By the expression, *off-the-shelf database systems*, Wright (2001b, p. 576) understands general purpose databases such as Microsoft Access that can be used for managing terminological data after configuring the database to meet terminology management needs. Database management systems offer powerful facilities for data modelling and retrieval and support creating link relations. General purpose databases are not ideal for linguistic data and they are not easy to handle. The layout of data categories in databases may be awkward to view as entries become very long. There are also limitations on the length of the strings of text that can be put in the data field. The user can arrange a convenient display of information on screen by application of forms, which provide a separate view for each entry and easy navigation between entries. The limitations on length can also be dealt with by changing the data type (for example from Text to Memo in Access). Off-the-shelf databases do not provide automatic consistency checks, concept orientation or term autonomy. However, the user can program these options easily. Defining the type of data to be used within the field prevents undesired types of information from appearing and contributes to the integrity of the information within the termbase. Concept orientation may be implemented by introducing data categories such as hyperonymy, holonymy, entity type, subject field that link a given concept with related concepts. Term autonomy is also possible as the user can input each term associated with a given concept (e.g. abbreviation, synonym) in an independednt term element within one terminological record.

Dedicated Terminology Management Systems (DTMSs) are database management systems that have been developed or configured specifically for the purpose of managing terminological data, e.g. MultiTerm, which is specifically oriented towards translation. Terminology management programs are exactly adjusted to terminology work. Their structure is definable, as the user specifies what data categories to include in the termbase, on what level in the organisation hierarchy they should occur (e.g. the entry level, the index level and the term level in MultiTerm) and what values they should have. The template of the terminological record is predefined by the software but may be adapted to the user's needs.

DTMSs provide powerful data modelling and retrieval. They also offer elaborated user management, convenient display of information on the screen, consistency procedures and interfaces to other applications such as translation memory without the need to convert data. Concept orientation and term autonomy are either provided or definable. The systems with fixed structures have predefined concept orientation and term autonomy, while in systems such as MultiTerm they have to be defined by the user.

Although MultiTerm seems to meet all the criteria a terminological management system should meet, it is, in fact, not completely suitable for the purpose of my project. Its structure is quite rigid when compared to MS Access. Once the termbase structure is established it cannot be easily changed, for example by changing an existing data category or deleting it. Such a process requires redefining the termbase structure and frequentlyinvolves loss of completed entries. Moreover, MultiTerm is primarily oriented towards translation and it presupposes one-to-one correspondence between terms. The solution I am looking for needs to focus more on terminology, particularly one cases when terms do not have equivalents in the target language or have more than one equivalent. Therefore, I decided that MS Access will better meet my expectations. An additional feature that works in favour of this solution is the fact that MS Access offers export facilities to XML files, which is the file format used by MultiTerm termbase. MultiTerm has the MT Convert tool that creates xdt file into which the xml file containing the actual entries can be imported. Thus, the conversion of the termbase in Access to the termbase in MultiTerm can be achieved if needed.

1.4.3 Design of the surveying termbases in MS Access

The flexible and definable structure of MS Access database which facilitatesthe addition and removal of data categories, change of data types within categories and the establishment of relations between different data categories at any stage of terminological work (as well as relatively straightforward conversion of MS files to an on-line dictionary), make this software package the best choice for my project.

I used MS Access to compile two monolingual termbases with foreign language equivalents: one for English terms and the second for Polish terms. Data categories in the English termbase include:

- ID
- Citation form
- Abbreviation
- Part of speech
- Subject field
- Definition
- Source of definition
- Examples (x 3)
- Source of examples
- Synonym
- Hyperonym
- Holonym
- Entity type
- Status
- Notes
- Author
- Date
- Equivalent

The Polish termbase looks very similar to the English termbase. The only difference between the two termbases is that the Polish termbase has one additional category, which is gender for noun and noun phrases.

The next step in the design of the termbases in MS Access after specifying data categories depends on selecting the type of data within data categories. Due to the fact that some fields, e.g. *subject field*, *part of speech* or *gender* are closed lists of items, it may be useful to create picklist for them. Automatic numbering is the obvious solution for *ID*, while the data category *date* requires a date setting with a predefined format for dates.

MS Access provides the required settings for these data categories. It enables setting the type of data for ID to auto number, date to date/time and subject field, part of speech, and grammatical gender to Lookup wizard. The data type for all other data categories is text. It is important to mention, however, that MS Access restricts the size of the fields with text to 255 characters, which is a constraint for such data categories as examples or a definition. The memo setting solves this problem as it allows an input size of up to 63,999 characters.

Specification of types of data supports the integrity and consistency of information within the termbases and contributes to automation of the process of data input. To advance this process even more, I defined the field sizes and data formats for different data categories. These settings are presented in Table 1-10.

It is important to indicate that the English termbase has two embedded termbases: one with subject field classification that includes the ten subfields of surveying as specified in 1.2.3 and one with parts of speech. The table with parts of speech includes the following items: noun, verb, adjective, adverb, noun phrase (NP), adjectival phrase (AP), verb phrase (VP), adverbial phrase (AP) and prepositional phrase (PP). The Polish termbase has an additional embedded table for gender. These embedded tables work as picklists in MS Access. The language used to fill in the Polish termbase is Polish, which means that subject field specification or grammatical information is provided in Polish.

The data type for sources has to be text. It cannot be a picklist because the list of sources I use expands each time I have to look for examples and definitions for terms outside the corpora. MS Access does not provide any format types for text as it does for date. I adopt the convention of providing short bibliographic entries to document sources of definitions and examples and I keep full references in the EndNote database.

Table 1-10 Settings of data categories in MS Access

Data category	Data type	Field size	Data format
ID	auto number	long integer incremented by 1	
citation form	text	100 characters	
abbreviation	text	100 characters	
part of speech	lookup wizard	embedded table: part of speech	
gender	lookup wizard	embedded table: gender	
subject field	lookup wizard	embedded table: subject field	
definition	memo		
source of definition	text	255 characters	
examples (1, 2 and 3)	memo		
sources of examples	text	255 characters	
synonym	text	255 characters	
hyperonym	text	255 characters	
holonym	text	255 characters	
entity type	text	255 characters	
status	memo		
notes	memo		
author	text	50 characters	
date	date/time	Short date	yyyy-mm-dd
equivalent	text	255 characters	

The input of data to MS Access may be a problem in databases that have many data categories such as the surveying termbases. I overcame this difficulty by the application of forms, which are generated on the basis of tables and allow users to see each record of the termbase in a separate window. The picklist buttons have been automatically transferred from tables to forms to automate the selection of repeated information such as subject field.

Figure 1-1 presents the form view for the entry *gon* from the English surveying termbase.

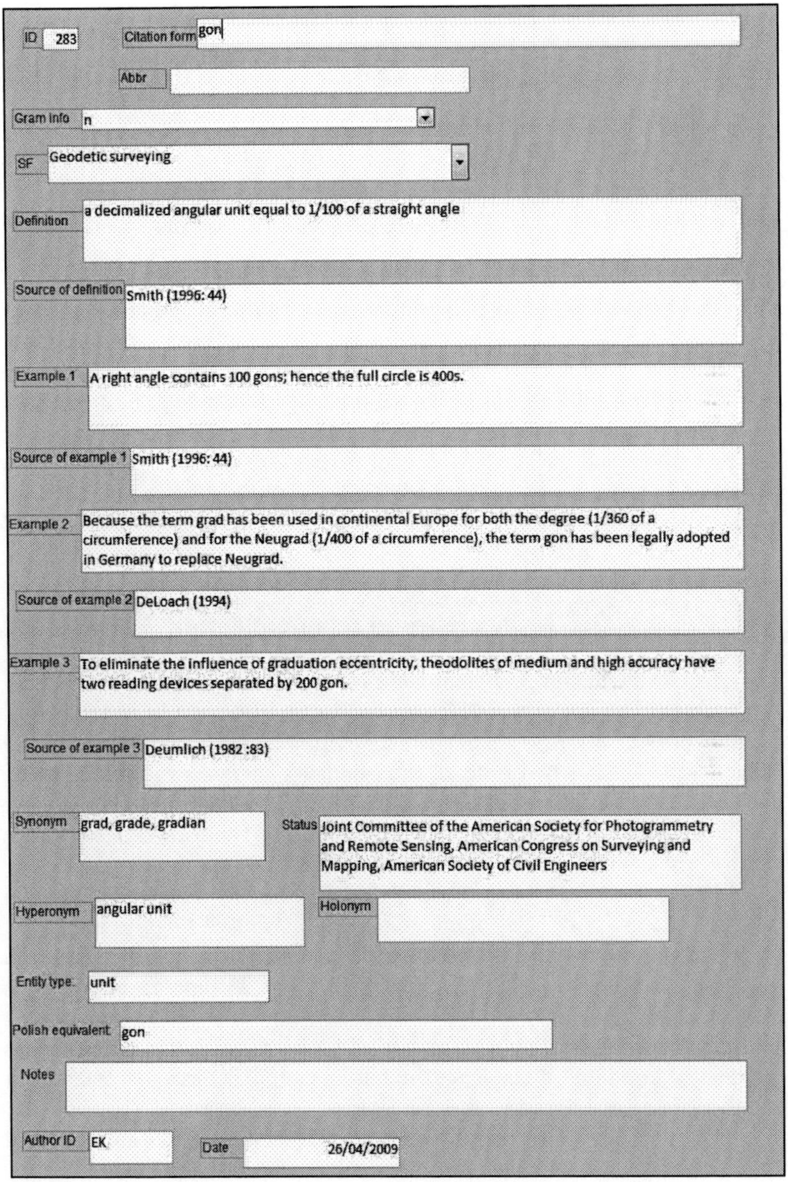

ID 283 Citation form gon

Abbr

Gram info n

SF Geodetic surveying

Definition a decimalized angular unit equal to 1/100 of a straight angle

Source of definition Smith (1996: 44)

Example 1 A right angle contains 100 gons; hence the full circle is 400s.

Source of example 1 Smith (1996: 44)

Example 2 Because the term grad has been used in continental Europe for both the degree (1/360 of a circumference) and for the Neugrad (1/400 of a circumference), the term gon has been legally adopted in Germany to replace Neugrad.

Source of example 2 DeLoach (1994)

Example 3 To eliminate the influence of graduation eccentricity, theodolites of medium and high accuracy have two reading devices separated by 200 gon.

Source of example 3 Deumlich (1982 :83)

Synonym grad, grade, gradian

Status Joint Committee of the American Society for Photogrammetry and Remote Sensing, American Congress on Surveying and Mapping, American Society of Civil Engineers

Hyperonym angular unit

Holonym

Entity type: unit

Polish equivalent gon

Notes

Author ID EK Date 26/04/2009

Figure 1-1 Termbase record from the MS Access surveying termbase in the form view

Chapter Two

Methodology of Term Collection

This chapter focuses on describing the procedure for collecting terms to be entered in the surveying termbase. First, different approaches to terminology compilation are presented and the role of corpora in terminology is outlined (2.1). Then the process of compiling the specialised corpora is described (2.2), with special attention being paid to the design of the corpora and technical aspects of corpus compilation. Next, the term extraction process is elaborated (2.3). Finally, the usefulness of concordancers and concordances in the compilation of term-related information is discussed (2.4).

2.1 Approaches to term collection

There is a wide range of approaches to terminology collection that may be adopted. They may be divided according to different sets of properties. Cabré (1999, p. 129) claims that terminological searches may be characterised using two criteria: the number of languages involved, and whether the search is systematic or not. The first criterion allows us to divide searches into monolingual and multilingual. The second differentiates between systematic and ad hoc searches.

Monolingual searches are usually carried out for lexicons of special subject fields and are often aimed at standardising terminology. Multilingual searches are used to compile dictionaries or vocabularies with information in several languages. Cabré (1999, p. 230) points out that it is useful to distinguish a third type of search within this category. She calls these searches 'monolingual searches with equivalents', and defines them as being carried out in one language but including a subsequent search for equivalents in one or more other languages.

The difference between systematic searches and ad hoc searches is that the former cover the terms of an entire subject field or one of its subfields, while the latter, which are typically text-based, are limited to a single term or a small set of terms that belong to a subfield of a subject field, or to a group of terms that belong to different fields. While terminologists using systematic searches collect terms designating particular concepts in a subject domain or subdomain in a structured way, those who use ad hoc searches try to address a specific problem or a terminological doubt of a user. Undertaking an ad hoc search is usually the result of a query that a user asks a terminological service. It involves three stages: the query, the search, and the response (Cabré, 1999, p. 153).

Wright (1997, p. 18) claims that terminology work can be either prescriptive or descriptive depending on the purpose. The aim of prescriptive terminology work is to indicate terms whose use is recommended (due to the fact that they are standard or standardised terms). The prescriptive work can be further subdivided into work that is strictly perspective about the forms of the terms, and work that is prescriptive about concepts. The prescriptive approach is used by standard bodies that lay down and define concepts, terms and equivalents, and classify them as preferred, admitted and deprecated. The objective of descriptive terminology work is to document all terms used to designate the concepts treated in a single discipline. Descriptive research is a basis for prescriptive work.

Sager (2001, p. 762) points out that terminology compilation is now adopting a corpus-based approach. The corpus-based approach to terminology originated in the early 1960s, when the first electronic corpus of English - the Brown Corpus - was developed. Brown was a collection of written texts in American English. Its British English equivalent – the Lancaster-Oslo-Bergen Corpus - was created a decade later (Atkins & Rundell, 2008, p. 58). The first attempts to capture terminology from electronic corpora were undertaken in the late 1980s (Ahmad & Rogers, 2001, p. 727).

The advent of electronic corpora means that both systematic and ad hoc term processing are now solidly corpus-based (Sager, 2001, p. 762). There are some situations, such as when electronic collections of texts are not available, that, in order to answer a specific query, the terminologist has to search the existing literature or needs to consult specialists. However, most terminological work is nowadays based on machine-readable corpora.

Sager (1990, p. 56) distinguishes between onomasiological and semasiological approaches to terminology work. The onomasiological approach starts from concepts and looks for the names of these concepts, while the semasiological approach starts from words and looks for their meaning. The onomasiological approach is still considered to be the most systematic approach to terms (Cabré, 1999, p. 8). According to Cabré (1999, p. 162) terminologists should typically start their work from concepts. In practice, however, terminologists typically start their work from a list of terms in a specific field which is extracted from the electronic corpus and combine this approach with the onomasiological one. Thus, the first thing they have is a list of words that constitutes the inventory of entries for a termbase or a dictionary. The terminologist describes them semantically by means of definitions. When writing definitions, terminologists become aware of concepts which are represented by terms, and semantic relations between concepts, such as hyperonymy or holonymy. This part of their work is semasiological as they start from forms and look for their meanings. Then, they come across concepts for which they have to find designations or instances when they have to choose between different designations for a particular concept and reject others, or accept one as the main form and others as synonyms. This part of their work is onomasiological as they proceed from concepts to their names (terms).

Van der Vliet (2006, p. 62) suggests that a system of concepts describing domain knowledge should be built by combining a top-down approach, which uses a domain knowledge, and a bottom-up approach that uses a corpus. The top-down approach relies on domain knowledge. The terminologist should acquire knowledge of a particular domain by reading handbooks, consulting experts, etc. A good understanding of a particular domain is the starting point for building a concept system for the particular field. Concepts need to be lexicalised as terms, so the terminologist has to use a corpus of relevant texts to extract candidate terms that can be linked to concepts. It may happen, however, that some candidate terms found in the corpus will not correspond to concepts that were identified using the top-down procedure. In such cases, a bottom-up approach needs to be applied. This approach uses a corpus as a source of data, and allows the terminologist to find terms and observe how they combine with other terms in compounds, collocations and sentences. Thus, the terms and their combinatoric properties may be a basis for collecting and structuring the domain knowledge, while the domain knowledge is a basis for finding terms.

The selection of approaches that have been used to perform term extraction is described in section (2.3).

2.2 Creation of the corpus

This section deals with three major aspects of corpus creation. First, the design of the corpus is considered (2.2.1), paying attention to the criteria for the design of the corpus and their application to the development of specialised surveying corpora. Then, the methods of data collection are discussed (2.2.2).

2.2.1 Design of the corpus

The advent of electronic corpora and term extraction systems changed the definition of what constitutes a corpus. It is nowadays understood as any collection of running texts held in electronic form and analysable automatically or semi-automatically rather than manually (Kurcz et al., 1990, p. 226). The corpus plays different roles. Ahmad and Rogers (2001, p. 740) claim that it may be used to capture data (e.g. to extract candidate terms), to validate data (e.g. to provide further evidence for or about a term candidate) or to elaborate data (e.g. to establish relations of synonymy, to find definitions or examples).

There are two main approaches to corpus: one which is specific to corpus linguists who treat a corpus as a collection of texts and another one, which is typical of mentalist linguists, who consider a corpus as a source of data to study underlying linguistic system that is in the mind of the speaker.

Corpus design is a complex process. Atkins and Rundell (2008, p. 54) claim that a perfect corpus for lexicography does not exist and cannot be accomplished. The corpus is only a sample of all the communicative events of a language. It is not possible to collect every instance of the use of any particular modern language as modern languages are used actively and undergo changes all the time. The only types of corpora that provide a complete record of evidence and therefore represent the language are corpora of historical languages such as Ancient Greek or Old English.

This last statement is controversial. It is true only if language is defined as a performance of a precisely delineated speech community. Such a definition is valid if a corpus linguistic view is adopted. This view considers language (performance) a corpus and which interprets a corpus as a collection of texts. There is an alternative approach to a corpus, however. It is called a mentalistic approach and treats a corpus as a source of data to study underlying linguistic system that is in the mind of the speaker. These two approaches have consequences in corpus design. Representativeness is central to corpus linguists, while it is not considered by cognitive linguists as there is nothing a corpus can be representative of.

Atkins and Rundell (2008, p. 57) claim that when designing a corpus, corpus builders need to focus on two main aspects: corpus size and corpus content. Corpus size is no longer a problematic issue. Ahmad and Rogers (2001, p. 735) point out that corpus builders are no longer constrained by the availability of computer memories and processing capacity, therefore they are able to collect as large a corpus as they need or can afford to create. Atkins and Rundell (2008, p. 61) claim that there is no definite minimum size for a corpus. However, they are aware of the consequences of frequency observations made by Zipf (1902-1950), who found that “a few words occur with very high frequency while many words occur but rarely”. This observation indicates that in order to get adequate information for the rarer words and rarer usages very large amounts of text are needed. A large corpus provides more evidence for linguistic features such as word combination or meaning, therefore the larger the corpus, the better.

When deciding on corpus content, corpus linguists have to strive for a corpus that represents the language (performance) of a precisely delimited speech community. Therefore they have to aim at obtaining samples of exemplar texts from a set that constitutes a universe of texts (Ahmad & Rogers, 2001, p. 734). Leech (1991, p. 27) claims that a corpus is representative to the extent that findings based on its contents can be generalised to a larger hypothetical corpus. Inferences about the language (performance) are thus made on the basis of a sample. In order to avoid a bias a ‘random sample’ should be collected. A random sample is one in which every member of the broader population has the same chance of being selected (Atkins & Rundell, 2008, p. 64). In other words, a random sample from the corpus should facilitate making generalisations about the performance in question.

There is no obvious way of creating a representative corpus of living languages as it is impossible to define the population that the corpus should represent. This population is unlimited. Since languages keep growing and changing, it is not possible to establish the correct proportions of each component (Atkins & Rundell, 2008, p. 66).

While the representative corpus is an unattainable goal, a ‘balanced corpus’ can be achieved. The aim of a balanced corpus is to reflect the diversity of language by including a full repertoire of ways in which people use the language. There are no scientific methods of obtaining a balanced corpus, as it involves many subjective decisions. Atkins and Rundell (2008, p. 75) list a number of text attributes that need to be considered when an attempt is made to create a balanced corpus. These attributes include:

- Authorship, i.e. whether the text was produced by one or more authors, whether the author(s) is male or female.

- Preparedness, i.e. whether the text is spontaneous, based on notes or fully edited.
- Function, i.e. whether the text is narrative, informative, expository, and persuasive.
- Audience, i.e. whether the text is aimed at adults, children, teenagers.
- Technicality, i.e. whether the text has been produced by specialists for specialists or whether it was produced by specialists and is directed at people or whether it is written for non-specialists.
- Mode, which refers to the type of material. A corpus may include written texts, spoken texts or a combination of both.
- Medium, referring to the channel in which the text appears. There are two basic types of media: print media and spoken media. Print media include newspapers, books, magazines, journals, dissertations, movie scripts, government documents and legal statutes. Spoken media consists mainly of face-to-face conversations, broadcasts and podcasts, public meetings and educational encounters, e.g. seminars, conferences.
- Domain, which refers to the subject matter of a text.
- Languages, i.e. whether corpus is monolingual, bilingual or multilingual. A distinction can be made between comparable corpora, which are collections of individual, monolingual corpora that contain completely different texts in several languages (McEnery & Wilson, 2006, p. 57), and parallel corpora, which are bi- or multilingual corpora that contain original texts and their translations in one or more languages (Teubert, 1996, p. 245). An example of a comparable corpus is the International Corpus of English, which consists of fifteen corpora representing different varieties of English, while the most prominent example of a parallel corpus is the English and French Canadian Hansard corpus (Atkins & Rundell, 2008, p. 70) or the JRC-Acquis corpus of the EU legal texts written between the 1950s and now and available in twenty-two languages (European Comission. Joint Research Centre, 2011).
- Time, which specifies from what period of time texts come. A synchronic corpus contains texts from a specific period of time, while a diachronic corpus consists of texts that come from an extended period.

Balance is very important in the corpus design. If a corpus fails to represent the diversity of style and content in the performance of a given speech community it is skewed (McEnery & Wilson, 2006, p. 386). Large corpora are less likely to be affected by skewing than small corpora.

While the design criteria for general-language corpora have been extensively discussed in corpus linguistics, less attention has been paid to the design of special language corpora. Ahmad and Rogers (2001, p. 736) claim that the criteria for general corpora design may in many cases apply to specialised corpora. Relevant criteria for special-language corpora include text type, domain, subdomain, language, regional variety, original or translated text, spoken or written mode, date, etc. The vocabulary used in special-language texts is much smaller than in general-language texts and a highly-specialised corpus of 100,000 words, would be a good starting point for terminology management (Ahmad & Rogers, 2001, p. 735).

When designing a surveying corpus I created two separate monolingual corpora: one in English and one in Polish. Since the field of surveying was analysed as consisting of 10 subfields, which have been listed in (1.2.3), I decided to select three of them: Geodetic surveying, GPS and Cartography for further analysis. For each of these subfields I created separate subcorpora, whose size varies from 35,000 to 45,000 words depending on the availability of materials for the corpus. The texts used to build the corpora come from different sources: technical magazines, journals, websites designed for surveying students and teachers, software and hardware manuals and textbooks. The size of the samples varies from 500 words for an article to 14,000 words for a section in a textbook or a journal article. The average size of text samples is between 3,000 words and 4,000 words.

Textbooks are terminologically denser than technical magazines because they are aimed at subject specialists, while magazines are designed for a wider audience, both specialists and non-specialists. The advantage of using a variety of texts is that it supports finding both new terms and standardised terms. New terms appear in the most recent publications in technical magazines before they are standardised and will be used in books.

The other aspects to be discussed when designing the corpus are the availability of texts and copyright. I am working with monolingual corpora and I find it relatively easy to find texts on surveying in English as there is an abundance of printed publications and electronic resources in this language. It is not so easy to find reliable resources for Polish. The selection of books is quite limited and many of them, especially these referring to more traditional subfields of surveying such as geodetic surveying, date back to the 1970s. There are useful websites but due to the fact that most surveying equipment and software are developed outside Poland, websites in Polish often contain only a fraction of the information which is included on the English websites. What is more, Polish textbooks rarely contain indexes which are very useful for terminological researches. For these reasons, the process of compiling a Polish corpus was more demanding and time consuming than compiling the English corpus.

Copyright is also an important consideration in building a corpus. I provide detailed information on the ownership of texts I have in the corpus both for printed and electronic texts. For this purpose I have created a separate database in EndNote, which provides only bibliographic references for the surveying corpora.

2.2.2 Collecting texts for the corpus

Since it became standard to produce corpora in electronic format, general-language corpora started to be increasingly available off-the-shelf. Conversely, special-language corpora are not readily available and they have to be compiled. They can rarely be re-used, as is the case for general-language corpora, because terminologies are domain-specific. Basically, terminologists have to build a new corpus each time they are involved in compiling a new terminology (Ahmad & Rogers, 2001, p. 732).

In order to build a corpus, texts need to be converted into machine-readable format, if they are not in this form already. Bowker (2002, p. 23) describes keying, scanning combined with optical-character recognition and voice-recognition technology as the main methods for converting a text into a machine-readable form. Ahmad and Rogers (2001, p. 733) indicate that the method that depends on re-using existing corpora that are available in machine-readable form should be added to this list.

The most common method of collecting electronic text depends on re-using texts which is already in machine-readable form. There is a wide range of texts available on the Internet, through such applications as electronic books (Google Books), electronic journals and magazines, traditional journals reproduced in electronic form, company and institutional websites, promotional materials, etc. Lexicographers view the web as a source of texts from which a corpus can be assembled (Atkins & Rundell, 2008, p. 78). Sources which are available in such file formats as DOC, RTF or TXT usually do not require any preparation and can be used for the corpus immediately. Those which are in PDF format have to be converted into one of the formats specified in order to facilitate term extraction.

The surveying corpora in my project were compiled through scanning and re-using materials which were already available in electronic format. The main sources of texts were textbooks, journals, technical magazines and websites. Most electronic texts could be copied directly from the websites and pasted into an MS Word document or a plain text file. However, quite a few journal articles were available in PDF format only, and required conversion to TXT and DOC formats, which can be used by concordance and term extraction software. The involvement of different methods of text acquisition Table 2-1.

Table 2-1 A quantitative analysis of methods of text acquisition used to compile surveying corpora in English and Polish

Meth-od of text acquisi-tion	English corpus				Polish corpus			
	In total	Carto-graphic corpus	Geode-tic survey-ing corpus	GPS cor-pus	In total	Carto-graphic corpus	Geode-tic survey-ing corpus	GPS cor-pus
PDF	34%	67%	13%	23%	6%	0%	17%	0%
Text	12%	7%	5%	23%	36%	32%	21%	55%
Scanning	54%	26%	81%	54%	58%	68%	62%	45%

Scanning was the main method of compiling texts as it provided me with 54% of texts for the English corpus and 58% of texts in the Polish corpus. The proportions of texts that were available as PDF files and which were already in machine-readable form are in inverse proportion in English and Polish. PDF files constitute 34% of the English corpus and only 6% of the Polish corpus, while texts which were already in machine-readable form make up 12% of the English corpus and 36% of the Polish corpus. This disproportion may be justified by the fact that the number of surveying journals which typically offer articles in PDF format is very limited in Polish when compared to English journals. Most texts in the Polish corpora come from textbooks and websites as they were the most readily available source of data.

All texts that create corpora were documented by providing each of them with a unique header that records its essential features. The headers of my input documents give short bibliographic information on the texts that have been used in the corpora. They typically provide the name of the author of the text and, if any specific corpus includes a few texts from a book or a website by the same author, they also include chapter numbers, section titles and date when the text was published. Short references may be linked easily to full bibliographic references in EndNote. Headers are very important not only in the identification of the text from which a particular term comes, but also in documenting sources of definitions and examples in the termbase.

2.3 Term extraction tools

According to Cabré et al. (2001, p. 53) the need to automatically extract terminological units from specialised texts arose in the late 1980s. The first large computerised text corpora were created in the 1990s and the process was followed by developing the first programs for term extraction.

The processing of texts using computer programs in order to identify strings that are potential terms is referred to as *automatic term extraction* or *semi-automatic term extraction* (Ahmad & Rogers, 2001, p. 725). Bowker and Pearson (2002, p. 165) claim that it is not correct to call this process automatic, as, although the initial extraction attempt is performed by a computer, the resulting list of candidate terms is just a proposal and actual terms have to be confirmed by a terminologist. Therefore, they support the view that the process should be called semi-automatic term extraction. Their reasoning seems to be justified, as Ahmad and Rogers indicate that term extraction produces the raw material for termbases which needs to be examined, tested and validated before inclusion in a termbase (Ahmad & Rogers, 2001, p. 725).

Bowker and Pearson (2002, p. 165) additionally make a distinction between monolingual and bilingual term extraction. Monolingual term extraction tools analyse specialised corpora in order to identify candidate terms, while bilingual term extraction tools analyse aligned bilingual corpora, trying to find candidate terms and their translation equivalents.

Cabré et al. (2001, p. 54) distinguish three systems for terminology extraction: a system with linguistic approach, a system with statistical approach and a system with hybrid approach, which is a combination of the linguistic and statistical approaches. In each approach systems analyse a corpus that consists of specialised texts in electronic form and extract lists of candidate terms which can either be accepted or rejected by the terminologist after analysing the context for candidate terms and other information such as frequency of occurrence, relationship between terms, etc.

The linguistic approach to terminology extraction depends on the identification of word combinations that match particular part-of-speech patterns (Bowker, 2002, p. 83). In English many terms are N+N and A+N compounds. In order to identify such terms, each word in the text has to be tagged with its appropriate part of speech. It can be done either manually by the user or automatically by the system if it is equipped with an automatic tagger. Once this has been done, the term extraction tool identifies all the occurrences that match these patterns.

Ahmad and Rogers (2001, p. 741) claim that the most straightforward statistical approach depends on a simple frequency count, in which each token (word form) in the corpus is listed according to its frequency of occurrence. Frequency lists include many single-word candidate terms and potential mother terms, which usually carry with them other terms and are used less frequently on their own. These single-word term candidates and mother terms feature a very high cross-disciplinary productivity in the formation of compounds or in the production of derivational forms. Ahmad and Rogers (2001, p. 741) make a distinction between closed class words and open class words. A closed class is a class of words to which no new items can normally be added. It includes a small number of items such as determiners, pronouns, conjunctions and prepositions. An open class, in contrast, offers possibilities for expansion. Typical open classes are nouns and verbs which may acquire new words by word formation processes. Ahmad and Rogers (2001, p. 742) point out that the first 100 most frequent open class words include a high proportion of mother terms. Closed class words should not be taken into account and the noise caused by them should be filtered. However, they may occur in multi-word units and disregarding them affects the extraction of multi-word units.

Bowker (2002, p. 84) explains the principles of the statistical approach in a slightly different way. She states that the tool in the most straightforward statistical approach will look for repeated series of lexical items. The user can often specify the frequency threshold, which means that, if it is set for example at two, a given series of lexical items must repeat at least twice in the text to be recognised as a term candidate by the term extraction tool. One drawback of this approach is that the software often extracts items that are not candidate terms as there are many repetitions in a corpus, and not all repeated series of lexical items are terms or candidate terms. Some of them constitute noise and have to be eliminated. Another drawback of this approach is that not all of the candidate terms that appear in the text will be repeated, which leads to silence. This approach, however, also has positive features. It is not language dependent which means that it can be used to process texts in various languages.

The third approach to term extraction, the hybrid approach combines both statistical and linguistic methods. Hybrid systems extract terms in two steps. The first step depends on extracting syntactically homogeneous material by means of symbolic extraction and the second step is some type of filtering, typically based on statistical measures for termhood or, by association, measures in the case of collocations (Heid, 2006, p. 104).

Both Ahmad and Rogers (2001, p. 751-752) and Manning and Schütze (1999, p. 268) claim that the performance of the extraction system should be measured against the four logical predicates presented in Figure 2-1, which include the correct identification of terms (cell A), the correct omission of non-terms (B), the incorrect omission of a term (C), and the incorrect identification of a non-term as a term (D). Cell D also covers partial hits, e.g. *electromagnetic distance* for *electromagnetic distance measurement*.

correct	incorrect	
term A	non-term (noise) D	identified
B non-term	C term (silence)	omitted

Figure 2-1 A schematic presentation of the four logical possibilities for hits and misses in term extraction after Ahmad and Rogers (2001, p. 752)

Terms that have been correctly identified (so called "true positives or tp" in A) and non-terms that have been omitted (so called "true negatives or tn" in B) do not pose problems to a term extraction system. Cases which are more difficult to handle include non-terms that have been incorrectly identified as terms (so called "false positives or fp" in D) and terms that have been omitted (so called "false negatives or fn" in C). False positives produce noise but it can be filtered out by either human intervention or linguistic rules. False negatives which produce silence constitute a potentially greater problem in term extraction. The four logical predicates are components of the most common measures used in the assessment of the efficiency of term extraction, which are precision and recall (Cabré, Estopà & Vivaldi, 2001, p. 54).

Manning and Schütze (1999, p. 268) define *precision* as a measure of the proportion of selected items that the system got right, and *recall* as a measure of the proportion of the target items that the system selected. In other words, precision is the capacity to discriminate between those units which are terms and those which are not, and recall is the capacity of the system to extract all terms from a document.

Taking into account that precision is defined as a measure of the proportion of selected items (*tp+fp*) the system got right, it may be expressed with the following equation:

$$precision = \frac{tp}{tp + fp}$$

Recall is the proportion of the target items (*tp*+*fn*) that the system has selected and is expressed with the equation:

$$recall = \frac{tp}{tp + fn}$$

Typically precision and recall may give contradictory information, e.g. precision may go down while recall goes up and thus may not be very useful in choosing an optimal noise level in term extraction. If this is the case, Manning and Schütze (1999, p. 269) propose the *F-measure* as a solution to this problem. The F-measure combines precision and recall into a single value. It is defined as follows:

$$F = \frac{1}{\alpha\frac{1}{P} + (1 - \alpha)\frac{1}{R}}$$

P is precision, *R* is recall and α is a factor which determines the weighting of precision and recall. A value of α =0.5 is often chosen for equal weigthing of precision and recall. With such a value, the F measure simplifies to 2*PR*/(*R*+*P*).

All approaches to term extraction that have been discussed so far seem to face the problem of extracting too much noise. This problem may be handled by the application of filters such as stop lists and exclude lists (Ahmad & Rogers, 2001, p. 742). Heid (2006, p. 103) points out that apart from typically lexical filters such as stop lists, there are also filters which are dependent on the outcome of statistical analysis. These filters accept combinations of a given syntactic form only if these combinations include at least one item which has been included in the list of candidate terms. The latter types of filter are used in the systems for term extraction that are based on hybrid approach.

A stop list or stop word list is a list of words which are filtered out, prior to or after processing of natural language data (text). The term was introduced by Hans Peter Luhn (1896-1964), one of the pioneers in information retrieval (Mani & Maybury, 1999, p. 13). Stop word lists are language-specific, which means they have to be created for each language individually, and typically contain articles, pronouns, prepositions, conjunctions and some highly frequent adjectives and adverbs (Heid, 2006, p. 103).

An exclude list is a list of items which need to be excluded from the term extraction process as they have a very low probability of being terms (SDL Trados, 2007, p. 324). An exclude list may be a basic vocabulary list. It usually includes frequent verbs, nouns, adjectives, adverbs, pronouns, prepositions, conjunctions as well as geographical names, nationalities, proper names, titles, days, months and their abbreviations, etc. Basic vocabulary is compiled taking into consideration the frequency of general vocabulary known to an average speaker of the language (SDL Trados).

In my project I used SDL MultiTerm, which is one of the two extraction tools offered by SDL. The other tool is SDL Phrase Finder. The basic difference between these two tools lies in the method of term extraction. SDL MultiTerm uses a statistical extraction method to determine the frequency of appearance of candidate terms, while SDL Phrase Finder uses a linguistic method for term extraction (SDL MultiTerm Extract: Tools Guide 2007).

Before I decided on a particular tool, I examined SDL MultiTerm Extract and SDL PhraseFinder in terms of such features as processing, supported languages, supported file formats and obviously the aforementioned extraction method.

SDL MultiTerm Extract uses a statistical method to determine the frequency of the appearance of candidate terms (SDL MultiTerm Extract: Tools Guide 2007). It extracts candidate terms and, for multilingual termbases, also their probable translations found in sentences, incomplete sentence fragments and strings of code. Statistical extraction is suitable both for small and large document processing and is quicker than extraction with SDL PhraseFinder. After running the extraction, SDL MultiTerm Extract displays a candidate term list and a terminologist has to validate candidate terms. SDL MultiTerm Extract has different export definitions. Terms can be exported to MultiTerm termbase, MultiTerm XML format and tab-delimited format.

SDL MultiTerm Extract provides filters in the form of stop word lists and basic vocabulary for a number of languages. It has stop word lists for seventeen languages. The stop word list for English has 392 entries. Unfortunately, no stop word list for Polish has been provided. SDL MultiTerm Extract provides basic vocabulary for five languages: English, French, German, Spanish and Italian. The English inventory has 4,279 entries. No basic vocabulary, however, is provided for Polish.

SDL MultiTerm Extract supports term extraction from a wide range of file formats. The supported file formats depend on the type of project. The monolingual file formats encompass such formats as plain text, rich text format, Microsoft Office: Word, Excel, PowerPoint; HTML, SGML, and XML, whereas the bilingual file formats available in the bilingual term extraction project, translation project and dictionary compilation project cover such formats as Translator's Workbench (TMW) and Translation Memory Exchange Format (TMX).

SDL PhraseFinder uses a linguistic extraction method to extract terminology candidates and is designed for small document processing. It uses intelligent algorithms, language rules and a lexical database to analyse the content of selected files and identify words and phrases. The candidate terms are displayed in the SDL PhraseFinder screen where a terminologist can view and validate terms. SDL PhraseFinder can analyse monolingual files and extract candidate terms from them, and bilingual files from which it extracts both the source language terms and their translations.

Multiple filters can be applied to ensure accurate extraction. They involve lexical filters, such as 'ignore capitalization' (with options to ignore first uppercase except start of sentence, all uppercase and duplicate terms with different capitalization), 'ignore numbers', 'ignore function words', such as articles, conjunctions and prepositions. The more sophisticated filters may set the software to ignore 'blocked lists', which contain words and phrases that the user wants SDLPhraseFinder to 'block' (not display) if they are present in the analysis. Blocked lists may include terms that have already been exported. Other filters are 'non-maximal matches', which exclude strings properly included in another candidate term, and 'unfound words in the database', which are words that do not appear in SDLPhraseFinder's lexical database. The user can also specify the maximum number of words per term.

SDL PhraseFinder only supports such languages as English, French, German, Spanish, Italian, Dutch and Portuguese. It is not possible to use it for Polish as this language cannot be selected from the list of source languages of the files that have to be analysed and, as a result, it is not possible to proceed to the next step in the extraction.

The software supports only a few file formats including TXT, RTF, HTML, Translation Document (ITD) and Translation Memory (MDB). This method takes more memory and computing takes longer than in the case of SDL MultiTerm Extract.

I considered all the features of SDL MultiTerm Extract and SDLPhraseFinder and came to the conclusion that the former better suits my purpose as it supports both English and Polish. It is more economic to use the same software package for two languages as I get a better level of comparison. Additionally, when I use the same software for different languages I only need to learn how to adjust settings in one programme.

SDL MultiTerm Extract supports five types of projects that may be created in MultiTerm Extract: Monolingual Term Extraction Project, Bilingual Term Extraction Project, Dictionary Compilation Project, QA Project and Translation Project. As I do not have access to suitable translation corpora in the domain of surveying and do not rely on existing termbases, only Monolingual Term Extraction Project may be selected in my case.

SDL MultiTerm Extract offers a wide range of settings which, if appropriately adjusted, can increase the efficiency of the selection of candidate terms. The range of settings available in a Monolingual Term Extraction Project which I have selected is listed in (7).

(7) a. Minimum term length
b. Maximum term length
c. Maximum number of extracted terms
d. Silence/noise ratio
e. Stopword lists
f. Exclusion files

Before I started term extraction I tested various settings of MultiTerm on small, comparable text corpora taken from English and Polish surveying textbooks. The purpose of this research was to find the settings that provide the highest proportion of terms in the list of candidate terms.

In my experiments I adopted default minimum and maximum term lengths (7a-b) and considered any successful occurrence of the term as a 'hit', for example when actual terms are parts of longer phrases. I did not use (7c) as I was interested in the whole set of candidate terms. However, I focused mainly on exploring the relation between settings (7d-f) and a number of candidate terms extracted from corpora.

In order to establish a benchmark for my experiments, I started by manually extracting the terms from the two corpora. In the experiments, I investigated the relation between noise ratio and number of candidate terms extracted from corpora as well as the inclusion of terms in the list of candidate terms. In order to evaluate extraction outcomes, I used precision and recall measures.

Before I move to a discussion of experiments, I will briefly describe settings (7d-e). Setting (7d) has the form of a scale with maximal noise (minimal silence) indicated by the leftmost point on the scale, maximal silence (minimal noise) indicated by the rightmost point on the scale and the default setting just in the middle of the scale as shown in the *New Project Wizard* window in Figure 2-2.

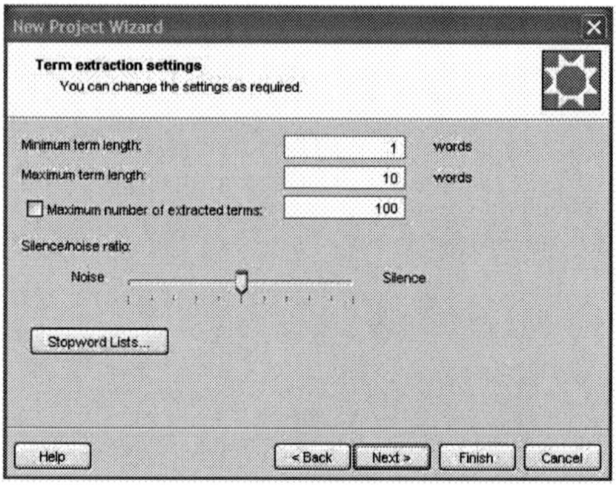

Figure 2-2 Default settings in MultiTerm Extract

Following Fernandéz Parra and ten Hacken (2008) I referred to this setting as *noise levels*, ranging from 0 (minimal noise) to 1 (minimal silence). The scale in Figure 2-2 indicates nine intermediate points between 0 and 1. The outcomes of term extraction proved, however, that the same output was obtained for noise levels from 0.8 to 1 and from 0.6 to 0.7. Taking this into account, there are eight noise levels to choose from. I also observed that candidate terms identified for the highest levels of noise are often very long. This is due to the fact that whole sentences are extracted.

Setting (7e) is used to specify a stop word list. The stop word list is activated in the setup of an extraction project. It can be deactivated if required. It is possible to edit stop word lists or to select other files. SDL MultiTerm Extract provides stop word lists for English but as no stop word list is provided for Polish I had to compile it myself. After analysing a few stop word lists available on the Internet, I decided that a lexeme frequency list for the PWN demonstration corpus (PWN, 2009b) would be the most reliable solution. It was elaborated for the purpose of the Polish PWN dictionary. The frequency list has 571 word forms, which are grouped under the most frequent lexemes. The lexemes on this list have a very high frequency and they cover all case forms for auxiliary and modal verbs, pronouns and high-frequency nouns such as *mężczyzna* ('man'), *praca* ('job'), or *rok* ('year'). For example, there are 49 word forms for the lexeme *być* ('be') and 22 word forms for the lexeme *on* ('he'). The frequency of the lexemes had been calculated by summing up the frequency of all forms with reference to the proportions of their occurrence in the Polish frequency dictionary *SFPW* (Kurcz *et al.*, 1990).

The final option, (7f), is not visible in Figure 2-2 but can be specified elsewhere in the project setup. Basic vocabulary is another device used in improving term extraction. SDL MultiTerm Extract provides basic vocabulary for five languages: English, French, German, Spanish and Italian. The English inventory has 4,279 entries. No basic vocabulary, however, is provided for Polish. Therefore, I had to compile a basic vocabulary for Polish myself, which proved to be very challenging. There are a few Polish corpora that could be the basis for creating a basic vocabulary file. They include the *IPI PAN* corpus (Przepiórkowski, 2008), the corpus of the *SFPW* (Kurcz *et al.*, 1990) and the *PWN corpus of Polish* (PWN, 2009a) . However, the type of access they give is constrained to a simple search. There are no corpora or corpora samples in Polish available to the public which could be used as the basis for a full concordance that would result in producing a basic vocabulary list. In these circumstances I had to look for other providers of basic vocabulary for Polish. I managed to find the basic vocabulary list for learners of Polish as a second language available at *Dicts.info* website (n.d). This basic vocabulary file has 1,514 entries from such categories as linguistics, mathematics, science, geography, weather, communication, nouns, adjectives, verbs, prepositions, etc. The interesting fact about this source of basic vocabulary is that it contains the same basic vocabulary for Polish, English and German so I assume that the basic vocabulary was created just for one language, probably English, and translated into the other languages.

For my experiments, I decided to use corpora that are small enough for manual extraction and for the evaluation of a number of experiments with alternative settings and at the same time dense on terms. Therefore, I decided to take fragments from textbooks in the domain of surveying. For English, I chose Bannister et al. (1998) as the textbook and selected the first two chapters as my corpus. The first chapter, entitled "Introductory", is 3,337 words long and the second chapter, entitled "Tape and offset surveying", is 2,454 words long. The total size of the corpus is 5,791 words.

Before automatic extraction, I identified terms in the corpus manually in order to obtain the target set. The target set consists of 120 terms. Then I carried out 32 term extraction projects for the English corpus. I performed automatic extraction with MultiTerm Extract for four sets of projects with 8 different noise settings. These sets of projects are characterised as in (8).

(8) a. Both the stop word list and the basic vocabulary are used
b. Only the basic vocabulary is used, no stop word list
c. Only the stop word list is used, no basic vocabulary
d. Neither the basic vocabulary nor the stop word list are used

Each experiment is characterised by one of the filter settings in (8) and a particular noise level. The result of each experiment is a list of candidate terms. The number of candidate terms in this list, which includes both terms and non-terms that have been selected is a first measure characterising the output of the projects. I had to analyse list of candidate terms and validate these items which are actually terms, which allowed me to establish values of *tp* (candidate terms extracted by the software which which are actual terms) and *fp* (candidate terms extracted by the software which are not terms), which are necessary to calculate precision. By comparing the list of terms extracted manually with the list of candidate terms extracted by the software, I could establish how many actual terms have been omitted by the term extraction system (*fn*) and thus calculate recall.

The extraction procedure for the Polish corpus followed the same pattern as applied to the English corpus. The composition of the Polish corpus was similar to the English one as I compiled it from a Polish textbook on surveying by Wołłodko (1973). This choice was not accidental as I attempted to achieve a correspondence with the English corpus, in order to increase the scope for comparison of the results. I selected fragments of the following chapters from the surveying textbook:

- Chapter 1 "*Podstawowe wiadomości z teorii błędów*" ('Basic information on the errors theory', 300 words).
- Chapter 2 "*Pomiary sytuacyjne*" ('Geodetic surveys', 1,451 words).
- Chapter 5 "*Niwelacja techniczna*" ('Technical levelling', 3,139 words).
- Chapter 7 "*Pomiar kątów*" ('Angle measurement', 760 words).

In the listed chapters, I only selected those parts that are dense in terms and contribute new information to the corpus, omitting tables and some passages with detailed descriptions of measurement processes. The total size of the Polish corpus created by applying such criteria reaches 5,650 words. In manual extraction I identified 139 terms.

The results of experiments proved that the activation of both the stop word list and the basic vocabulary for the English corpus eliminates a great amount of noise compared to projects where these filters are not activated. The tendency for Polish is similar, but compared to English, the basic vocabulary list has less impact on the set of candidate terms returned. I assume that if I had access to a better list or could devise one; for instance by analysis of Kurcz et al. (1990), the results for Polish would be more similar to the results obtained for English.

In both languages I found that when both filters are activated the increase of noise levels results in a rise of recall and a fall in precision. However, precision is much higher in English than in Polish, while the tendency for recall is opposite at all noise levels except for 0.8 to 1. Using Manning and Schütze's (1999) F-measure, I found that for both languages the rise in recall is much more significant than the fall in precision. Nevertheless, given the amount of manual validation work involved, there is a good case to be made for not using these highest noise levels. Even for small corpora, around 1000 individual candidates have to be analysed. In English, the case for a lower noise level is stronger, because at 0.6-0.7 precision is much higher and the number of candidate terms drops by almost 80%. For Polish the effect is less striking, which may be caused by the poor quality of the basic vocabulary. A detailed discussion of individual experiments and results is found in Kwiatek & ten Hacken (2011).

The outcomes of the experiments indicate that I need to use both stop word list and basic vocabulary and adjust the noise level to 0.6-0.7 to increase the efficiency of term extraction system and limit the amount of manual work involved in term validation.

I used these settings to extract terms from the English and Polish corpora. Processing the whole surveying corpus of approx. 120,000 words either in English or Polish proved to be an impossible task for MultiTerm Extract as it encountered errors while reading files. Therefore, I decided to extract candidate terms from each sub-corpus individually, which could be accomplished by the software. The number of candidate terms extracted from the English corpora is three or four times lower than the number of candidate terms for Polish. This may be caused by the highly inflective nature of the Polish language, which results in many morphological variants of the same lemma, as well as the quality of the basic vocabulary list, which proves to be incomplete for Polish.

After extracting candidate terms I validated those candidate terms that were actually terms. I sorted the list of validated terms manually, deleted duplicates and provided canonical forms for terms if they were in other forms. Then I exported terms into plain text file format. The number of candidate terms and validated terms for the English subcorpora is presented in Table 2-2.

Table 2-2 Number of term candidates and number of actual terms in English corpora

	Number of candidate terms	Number of terms
Cartographic corpus	1,472	100
GPS corpus	2,371	137
Geodetic surveying corpus	1,688	162

The total number of terms in the English surveying corpus was 399. There were 43 terms that occurred in two or more sub-corpora resulting in approx. 11% overlap of terms. I deleted all duplicates and obtained a list of 356 terms in English. The total number of 490 entries in the English termbase was acquired by using the onomasiological approach after the semasiological one.

Term extraction procedure for Polish surveying terminology was very similar. It required more manual work, however, as Polish terms rarely occurred in their canonical forms and required changing endings so that they could be given in the nominative case and in a singular number. The number of candidate terms and validated terms for the Polish subcorpora is presented in Table 2-3.

Table 2-3 Number of term candidates and number of actual terms in Polish corpora

	Number of candidate terms	Number of terms
Cartographic corpus	6,640	132
GPS corpus	7,649	99
Geodetic surveying corpus	4,182	161

It is quite surprising for Polish data that the smallest proportion of terms was validated in the longest list of candidate terms. This discrepancy may be accounted for by the fact that it occurs in GPS terminology, where many terms are borrowed from English. These terms do not have standardised labels in Polish but are simply based on direct translation of English terms. They are often general language words which work as terms in the field of GPS.

The total number of terms in Polish is 392 with 51 terms that occur in more than one sub-corpus, which is interpreted as a 13% overlap. After removing duplicates, the number of terms in Polish in the three sub-disciplines equals 341. The number of terms was increased to 459 in the Polish termbase by using the onomasiological approach.

2.4 Concordances

Sinclair (1995, p. 32) defines a concordance as a "collection of occurrences of a wordform, each in its own textual environment". In other words, a concordance is an index to the words in a text. A concordance indexes each word-form from a text and gives a reference to the place in which it occurs within the text. Concordances allow users to compare different usages of the same word and to analyse word frequencies. They also support finding phrases and idioms, searching for translations of substantial elements, as well as creating indexes and word lists. While a concordance is an index of the words that occur in the text, a concordancer is a tool used to produce such an index. Concordancers are used extensively these days to create glossaries and dictionaries (Lamy & Klarskov Mortensen, 2011). The KWIC format (Key Word in Context) is widely used in concordances (Sinclair, 1995, p. 32). This format puts the word-form under examination in the centre of each line, with left and right context. The user determines the constraints such as the number of words to the left and to the right of the searched word-form, as well as ordering. A basic concordancer orders occurrences in text order, but for most purposes, such as finding contexts for a particular term, it is helpful to use alphabetical ordering to the right or left of the central word-form.

An important consideration to keep in mind when discussing concordancers is the type of corpus being used. Concordancers can operate both on a raw corpus and on annotated and marked-up corpora. An example of annotation is a linguistic annotation. A linguistically annotated corpus may be searched more narrowly, e.g. a terminologist may wish to retrieve all instances of the word *transfer* when it appears as a noun.

Concordancers may be used to work with one language only or may be used for processing bilingual or multilingual texts. Monolingual concordancers may work in two different ways. Some monolingual concordancers operate by searching the whole corpus every time a search pattern is entered. Others work by first creating an index of all words in the corpus, along with a record of the location of each occurrence (Bowker, 2002, p. 53). The index must be created before any searches are conducted in a given corpus, but once created it may be reused each time a search for a particular item is done. The full-text search does not require any pre-processing and the corpus can be easily modified by adding or eliminating texts, but the search may take quite a long time, especially in the case of a large corpus.

Indexed searching involves pre-processing but the search is very quick, even for a large corpus. The disadvantage of this method is that each time the corpus is modified a new index must be created. An example of a concordancer that performs indexed searches only is a WordSmith Concord, which is a component of WordSmith Tools developed by Mike Scott at the University of Liverpool, whereas full-text searches (in addition to indexed seacrhes) are possible in Concordance by R.J.C. Watt of Dundee University (Lamy & Klarskov Mortensen, 2011).

Bilingual and multilingual concordancers permit parallel processing of texts in different languages. The parallel concordancer finds the equivalent sentences in the translated text. The operator has to prepare the corpus for use with parallel concordancers carefully. The two (or more) texts must be aligned in advance paragraph by paragraph, so that paragraph 2 in one language is equivalent to paragraph 2 in the other (Lamy & Klarskov Mortensen, 2011). Alignment can also take place at the sentence level or at the level of the whole text, chunks of texts or even words (Bowker, 2002, p. 56). However, the most efficient way of alignment is alignment either at sentence level or paragraph level. Alignments at text level are not useful in finding an equivalent for a particular expression, while alignment at word level is too difficult and error-prone, as there may not be one-to-one correspondence between languages. Alignment is a complicated process and most concordancing software separates the process of alignment and bilingual concondance generation (Bowker, 2002, p. 56).

In my research I only used monolingual concordances. Parallel concordances would benefit my research as they make the search for equivalents easier. However, the number of publications on surveying in English that have been translated into Polish is very limited. Polish texts on surveying that have been translated into English are even scarcer. Certainly, there are not enough resources of this type to build corpora that could be used to make a parallel concordance.

Concordancers are available as stand-alone packages and as components of other packages. Examples of stand-alone packages include Concordance (version 3.3) and MonoConc. WordSmith Tools (version 5.0 and 6.0) also contain a Concord component. There are also concordancing applications that work on the web, e.g. WebCorp, operated by Birmingham City University, which relies on different search engines. It includes a word-list generator that produces a word frequency list. An example of the software package which includes concordancer is SDL MultiTerm Extract.

In my research I mainly used MultiTerm Extract, which already contains a concordancer. This concordancer was sufficient for my needs in the majority of cases as it retrieved relevant information. If, in the process of writing up terminological records I came across terms that were not gleaned in the term extraction process, I searched for their origin and context in the corpus using WordSmith Tools.

The WordSmith Tools package is an integrated suite of programs for looking at how words behave in a corpus (Scott, 2010b). The package is based on the MicroConcord concordancer and is available in 6 versions. Versions 1 to 4 were published by Oxford University Press. The 5.0 and 6.0 versions are available from Lexical Analysis Software Ltd (Scott, 2010a). The package includes such applications as Concord, WordList and KeyWords.

Concord, which is a concordancer, supports searching single terms as well as a list of terms. Concord presents a concordance display and provides information about collocates of the search word, dispersion plots showing where the search word came in each file, and cluster analyses showing repeated clusters of words (phrases) etc. (Scott, 2010b).

The WordList facility generates word lists based on one or more ASCII files. The word lists are automatically generated in both alphabetical and frequency order. Word lists can be used to study the type of vocabulary used, to identify common word clusters, to compare the frequency of a word in different types of texts, or to generate a list of key words.

KeyWords is a tool for identifying the "key" words, which are essential to understanding what the text is about, in one or more texts (Scott, 2010b). The frequency of these words is unusually high in comparison with a reference corpus. The program compares two pre-existing wordlists, which must be created using the WordList tool. One of these is assumed to be a large word-list which will act as a reference file. The other is the word-list based on a selected text to be examined.

WordSmith has a utility called Language Chooser that allows the user to select languages to be processed. The software makes a distinction between A languages, which are English, German, Russian, Dutch and Finnish, for which all the settings are adjusted automatically, and B languages, which can be added to the existing list of the five languages but which require a manual selection of the settings.

It is also important to note that languages vary considerably in their preferences regarding sorting order. In languages such as French, accented characters are by default treated as equivalent to their unaccented counterparts (e.g. in French the order is *donne*, *donné*, *données*, *donner*, *donnez*, etc.). However, in other languages accented characters are treated as separate characters with their own position in the alphabet (in Polish the order is *dom .. dzwon .. dźwięk .. dżem ..).*

The user may adjust different settings in WordSmith. Some of these settings refer to the three tools (Concord, WordList, KeyWords) and they include the use of colours and the format for displaying information, printing settings and tag settings. The other settings are specific and they refer directly to each of the three tools. For the Concord tool, the user can decide how much context is displayed. A KWIC condordance with default settings shows five words to the right and five words to the left of the searched word (Scott, 2010b). The sorting preference is also quite important as it defines how context will be displayed. For WordList the user can set the length of the phrase, word frequency and tags. For keywords, the reference corpus and statistical method for determining which words are key words in the text may be adjusted.

I used a corpus of nearly 16,000 words in English and examined how it was processed using WordSmith Tools (version 5.0). I manually extracted 60 terms from this corpus and saved them in a plain text file as it is the only type of the file that WordSmith Tools supports. I used this file as a list of the searched words. Concord performed a concordance very quickly and proposed a number of sorting patterns to select from. I selected the sorting pattern 'Centre' which put the searched words in the centre of the line and displayed searched words in alphabetical order. I also experimented with a wildcard sorting pattern in the case of single searched words. It worked well and provided results which were not found when the list of the searched words was entered to find occurrences for items from this list. In this way, the search for the term *signal* provided such results as *signal*, *signals* and *signal-processing,* while the basic search gave just *signal*.

I also tested WordSmith for a Polish corpus that consisted of nearly 10,000 words. For the Polish corpus individual wildcard searches were much more effective than searching the whole list of words. Wildcard searches enable all morphological forms of the same lexeme to be viewed along with their contexts, and thus are very useful in finding examples for the termbase. Polish is a highly inflected language. Polish nouns have seven cases of declension, both singular and plural, while declension of adjectives additionally varies by gender which may be masculine, feminine or neuter. Polish verbs have six inflected forms in present tense (three forms in singular and three in plural). The verb forms vary additionally depending on aspect, mood or voice. A more detailed discussion of Polish morphology is presented in section 3.1.1.

By investigating different features and settings of WordSmith Tools, I decided that the wildcard search facility in Concord is the most useful application for my single searches. It supports finding examples for the searched terms, and by displaying the range of contexts it is helpful in defining concepts and obtaining grammatical information on terms such as irregular plural forms, e.g. *axis* vs *axes*.

CHAPTER THREE

ANALYSIS OF TERMS

The aim of this chapter is to analyse how terms are created and how they are named. Section 3.1 elaborates on those linguistic processes which depend on word formation and section 3.2 discusses terminological processes which constitute the term-naming act. Section 3.3 provides the summary of the findings. The surveying termbase described in the course of previous chapters will be the basis of this analysis.

In the section on linguistic processes I investigate word formation processes individually, keeping in mind that many entries from the termbase are morphologically complex as they were created using more than one word formation process, e.g. *technika cross-correlation* 'cross-correlation technique'. The term is borrowed from English and it incorporates the English compound *cross-correlation. Cross-correlation* was created in English as a result of multiple affixation with the prefix *cross-* and the suffix *-ion.*

In the section on terminological processes I study how the actual terms are created, taking into account subsequent stages in the name giving act and the motivations behind them. I consider various name-giving processes which, in addition to word formation, cover borrowing and semantic change and consider where in the lexicalisation process they take place.

3.1 Linguistic processes

In this section I look at word forms in the English and Polish surveying termbases and analyse which word formation processes are involved in the creation of surveying terms. I also try to find the most common word formation tendencies in the terminological databases (TDBs) presented in chapter one. Section 3.1.1 is a brief introduction to Polish morphology and phonology, which is necessary to understand word formation processes in Polish. Section 3.1.2 discusses both morphological and non-morphological word formation processes. It starts with a concise introduction to morphology and its key issues, such as morpheme and stem, as well as information on the make-up of English and Polish words. Section 3.1.3 is devoted to analysing the creation of multi-word units, which constitutes the next step in terminology. The creation of multi-word units involves word formation but goes far beyond it, e.g. *Royal Institution of Chartered Surveyors* is a multi-word unit which involves a compound *Chartered Surveyors*.

The sub-sections on word formation have a two-fold structure. The first part is theoretical, containing a definition of the process, a discussion of its features and the categorisation of formations which represent a particular process. The second part of each section includes a qualitative and quantitative analysis of the particular word formation pattern based on data from the English and Polish surveying termbases. Both parts are illustrated with numerous English and Polish examples.

In this and next chapter I will adopt the convention of quoting Polish examples by giving a Polish word in italics, which is followed by a morpheme by morpheme gloss into English in single quotation marks and the English equivalent, e.g. *gwiazdozbiór* 'star collection', i.e. constellation.

3.1.1 Remarks on Polish phonology and morphology

Polish is a highly inflective language. It distinguishes between masculine, feminine and neuter gender classes and there are seven cases of noun and adjective declension (in Polish): nominative, genitive, dative, accusative, instrumental, locative and vocative. The gender class of the noun can usually be determined by the nominative singular ending. Most masculine nouns in Polish end with a consonant, e.g. *sygnał* 'signal', *obiektyw* 'lens', *teren* 'terrain'. These nouns have stems which are identical to their citation forms and do not have any inflectional ending in the nominative case. Most feminine nouns and neuter nouns consist of a stem and a final vowel, which is their inflectional ending: *-a* for feminine nouns and *-o* or *-e* for neuter nouns e.g. *map(a)* 'map', *komparacj(a)* 'comparison', *legend(a)* 'legend', *pudł(o)* 'box', *dzieck(o)* 'child', *przewyższeni(e)* 'height difference'. There are exceptions to this rule as some masculine nouns have feminine inflectional ending, e.g. *satelita* 'satellite', *kolega* 'male colleague'. Other nouns inflect like feminine nouns, although they are masculine e.g. *sędzia* 'judge', *starosta* 'prefect (of a district)', *mężczyzna* 'man'. In fact, *sędzia* and *starosta* name both masculine and feminine professions (Klemensiewicz, 1962, p. 43). Gender in linguistics is considered in terms of agreement between nouns and articles and adjectives, which have different endings to agree with the gender of the noun (Yule, 2010, p. 84). The gender class of the expressions may be established by looking at the gender of the adjective, e.g. *doświadczony sędzia* 'experienced judge', where the adjective ends in a *–y* vowel, which is typical of masculine adjectives in Polish. Abstract feminine nouns end in *-ość*, e.g. *długość* 'length', *wartość* 'value', *wysokość* 'height', and some neuter nouns end in *-e*, e.g. *odwzorowani(e)* 'projection', *zniekształceni(e)* 'distortion'. The declension of nouns ending in *-ość* shows that their stem is phonologically exactly identical to their citation form (e.g. *długości* GEN, *długości* DAT, długości-ą INSTR) and the only difference depends on the orthography *ć : ci*.

At this point, it will be useful to discuss briefly Polish phonology and orthography. Some sounds may be written differently although they are pronounced in the same way, e.g. *h* and *ch* as in *hotel* and *chmiel* 'hop', *ż* and *rz* as in *żona* 'wife' and *rzeka* 'river', *u* and *ó* as in *mucha* 'fly' and *góra* 'mountain'. Polish consonants may be divided depending on the position of the middle part of the tongue into hard consonants, which are pronounced with the middle part of the tongue in the neutral (lowered) position, as *k* in *kot* 'cat', or *r* in *ryba* 'fish', and soft or softened consonants, which are pronounced with the middle part of the tongue lifted towards the hard palate as in *ćma* 'moth' or *pies* 'dog' (Mizerski et al., 2000, p. 14). Soft consonants are spelt *ć, dź, ń, ś, ź*, whereas softened consonants are spelt *ci, dzi, ni, si, zi*. Their pronunciation may be written down using the International Phonetic Alphabet (IPA) symbols as /tɕ/, /dʑ/, /ɲ/, /ɕ/, /ʑ/. Consonants with an accent are used before a consonant or at the end of a word, e.g. *ćma* 'moth', *dźwig* 'crane', *koń* 'horse', *śliwka* 'plum', *źrebak* 'foal', whereas *ci, dzi, ni* etc. are used before vowels the *a, ą, e, ę, o, u*, e.g. *cień* 'shadow', *gniazdo* 'nest', *dzięcioł* 'woodpecker', and in some plural nouns, e.g. *dni* 'days', *dzieci* 'children'.

Polish verbs have two different stems depending on aspect: one which is perfective and describes an action that will be completed in the future as in *z-jeść* 'to eat', *na-rysować* 'to draw', and one which is imperfective and describes an action that lasts, e.g. *jeść* 'to eat', *rysować* 'to draw'. Verbs in the perfective aspect may typically be recognised from those in the imperfective aspect by the presence of prefixes e.g. *u-* as in *ugotować* 'to cook', *prze-* as in *prze-czytać* 'to read', *na-* as in *na-pisać* 'to write'. However, there are few verbs that have irregular aspect partners, e.g. *brać* (imperfective) 'to take' vs *wziąć* (perfective) 'to take', *dawać* (imperfective) 'to give' vs *dać* (perfective) 'to give' (Mizerski *et al.*, 2000, p. 98).

Polish verbs additionally have an inflectional ending that attaches to the stem. For example, the infinitive *siadać* 'to sit' with the stem *siada* (the same as 3rd person singular which has ø ending) inflects in the following way: *(ja) siada(m)* 'I sit', *(ty) siada(sz)* 'you sit', *(on) siada* 'he sits', *(my) siada(my)* 'we sit', *(wy) siada(cie)*, you sit, (oni) siada(ją) 'they sit'. In some cases the stem may be deduced from the infinitive form of the verb as in *siadać 'to sit'* or *śpiewać* 'to sing', where the stem is *śpiewa* 'he sings', but in many cases various alterations between the infinitive and the stem occur, e.g. the infinitive *pisać* 'to write' has the stem *pisz* as in *pisz(ę)* 'I write', *pisz(esz)* 'you write', *pisz(e)* 'he writes', or the imperfective infinitive *grzać* 'to heat' has the stem *grzej* as in *grzej(ę)* 'I heat', grzej(esz) 'you heat', grzej(e)'he heats'. These alterations may be even more explicit if both the derivational and inflectional processes which occur in the whole family of words are considered, e.g. *wóz* 'cart', *woźnica* 'coachman', *wozowania* 'coach garage', *wózek* 'trolley', *powóz* 'carriage', *wożę* 'I carry' (*wóz-, woź-, woz-, woż-)*. Klemensiewicz (1962, p. 43) gives the most common alterations between the stem of the base word and the derived word, which are the hard and soft consonant alterations listed in (9), the hard and functionally soft consonant alterations given in (10) and the vowel alterations in (11). Klemensiewicz uses colons to indicate the relation between stems in derived words and an apostrophe after the consonant to show that the consonant is softened by the vowel 'i'.

(9) *p : p'*, e.g. *chłop* 'adult man' : *chłopiec* 'boy'
b : b', e.g. *dąb* 'oak' : *dębina* 'oak trees'
m : m', e.g. *sam* 'himself' : *samiec* 'male'
w: w', e.g. *sława* 'fame' : *sławić* 'to glorify'
f : f', e.g. *harfa* 'harp' : *harfiarz* 'harpist'
d : dź, e.g. *rada* 'advice' : *radzić* 'to advise'
t : ć, e.g. *złoto* 'gold' : *złocić* 'to cover with gold'
z : ź, e.g. *koza* 'goat' : *koźlę* 'young goat'
s : ś, e.g. *kosa* 'scythe' : *kosić* 'to scythe'
n : ń, e.g. *pan* 'master' : *państwo* 'Mr and Mrs'
k : k', e.g. *okno* 'window' : *okienko* 'small window'
g : g', e.g. *gra* 'game' : *gierka* 'small game'
ch : ch', e.g. *słuchać* 'to listen to' : *podsłuchiwać* 'to eavesdrop'

(10) *r : rz*, e.g. *stary* 'old' : *starzec* 'old man'
ł : l, e.g. *mały* 'small' : *malec* 'small child'
ch : sz, e.g. *głuchy* 'deaf' : *głuszec* 'capercaillie'
d : dz, e.g. *rada* 'advice' : *radzenie* 'advising'
t : c, e.g. *zapłata* 'pay' : *płacenie* 'paying'
z : ż, e.g. *groza* 'dread' : *grożenie* 'threatening'

k : cz, e.g. *krzyk* 'shout' : *krzyczenie* 'shouting'
z : dż :, e.g. *jazda* 'ride' : *jeżdżenie* 'riding'

The consonants *rz, l, sz, dz, c, ż, cz, dż* are hard consonants, just like *r, ł, ch, d, t, z, k.* They occur in the same places in words and affect the pronunciation in a similar way to soft consonants, e.g. *radzić : rada* (dź :d) vs *radzenie : rada* (dz : d). They play the role of soft consonants and, for this reason, they are called *functionally soft consonants.*

(11) *e : o*, e.g. *żeni-ć się* 'to get married' : *żon-a* 'a wife'; *lecieć* 'to fly' *: lot* 'flight'
e : a, e.g. *mierzyć* 'to measure' : *miara* 'the measure'; *słyszeć* 'to hear' : *słyszał* 'he heard'
o : ó, e.g. *doł-ek* 'small pit': *dół* 'large pit'; *pomoc* 'help' : *pomóc* 'to help'; *wiodłem* 'I led' : *wiódł* 'he led'
ę : ą, e.g. *dęb-owy* 'of an oak tree' : *dąb* 'oak'; *sędzia* 'judge' : *sądzić* 'to judge'
e : ø (it means that the *e* vowel may appear or disappear and the root in the derived word will not change), e.g. *lew* 'lion' : *lwic*a 'lioness'; *len* 'linen' : *lniany* 'linen, made of linen'

3.1.2 Word formation processes

The aim of this section is to discuss various types of word formation process on the basis of data from the surveying termbases. First, a distinction is made between morphological processes and non-morphological processes, which include borrowing and coinage. Then, key issues in morphology and the make-up of words in English and Polish are scrutinised. This is followed by an in-depth analysis of each word formation process as far as is relevant to the termbases.

Morphological processes depend on the creation of new words by applying morphological rules. According to Bloomfield (1933, p. 207), morphological word formation processes encompass derivation and compounding. Plag (2003, p. 17) takes Bloomfield's classification as a starting point to differentiate between derivation with affixation and derivation without affixation. Derivation with affixation covers prefixation and suffixation, and non-affixational derivation incorporates conversion. Morphological word-formation processes also incorporate shortening processes and neoclassical word formation. Non-morphological processes rely on adapting words from other languages, which is known as borrowing, and the coinage of new words.

I will group word formation processes according to their features and discuss them in separate subsections in the following order: derivation (A), conversion (B), compounding (C), neoclassical word formation (D), shortening processes (E), analogy-based processes(F) which result in coinage of new words, and borrowing (G).

Before I start discussing morphological word formation processes, it is essential to introduce some key issues in morphology. Morphology refers on the one hand to word structure and on the other hand to the branch of linguistics which studies this (Trask, 1999, p. 194). Words can be analysed as having internal structure and consist of smaller units called morphemes. For example, the verb-form *talks* consists of two morphemes: the verb-stem *talk* and the grammatical ending *-s*. Such examples show that a broad distinction between two types of morpheme can be made. Morphemes can be divided into free morphemes, which can stand by themselves as single words, e.g. *talk* and bound morphemes which cannot normally stand alone and are typically attached to another form, e.g. *-s* (Yule, 2010, p. 63). All affixes are by definition bound morphemes. Free morphemes include separate word forms such as basic nouns, adjectives, verbs, etc. Some bound morphemes, for example *un-* are attached before the central meaningful element of the word, the so-called stem, while other bound morphemes, such as *-ed* follow the stem. The element preceding the stem is called a prefix and the element which is attached after the stem is the suffix. For example:

un-	*interest*	*-ed*
prefix	stem	suffix
(bound)	(free)	(bound)

Apart from those prefixes which are bound morphemes, there are also bound stems, which are stems that occur only in combination with other bound morphemes (Plag, 2003, p. 10). In English, many bound roots are of Latin origin, e.g. *circul-* (as in *circulate, circulation, circulatory, circular*) or *simul-* (as in *simulate, simulation, simulant*). These combining forms of neoclassical compounds which may occur both in the initial and final position in the compound are also called bound stems, e.g. *philo-* as in *philology* and *-phile* as in *bibliophile*. Polish verb stems can also be classified as bound stems as they only occur in combination with inflectional endings, e.g. *pisać* 'to write', which has the stem *pisz-* (as in *pisz-ę* 'I write', *pisz-esz* 'you write', *pisz-e* 'she writes', *pisz-emy* 'we write', *pisz-ecie* 'you write', *pisz-ą* 'they write').

Morphology is divided into two main areas. These are inflection, which concerns the variation in form of a single word for grammatical purposes, as with *talk*, *talks*, *talked*, *talking,* and word formation, which involves the construction of new word forms from existing words, as with *handbook* from *hand* and *book* and *player* from *play* and *-er* (Trask, 1999, p. 194). I will make a distinction between derivational endings and inflectional endings by hyphenating the former and bracketing the latter, e.g. *-er* as in *receiver, (ed)* as in *walked.*

In this thesis, I will concentrate only on those word formation processes which, unlike inflection, lead to the generation of new words and terms and therefore are important in the analysis of terms.

A. Derivation

Derivation relies on affixation. There are two main types of affixation processes:

- Prefixation, which depends on adding affixes to the beginning of the word, e.g. ***re**measure*, ***para**krok* 'double step'. These affixes are called prefixes.
- Suffixation, which depends on adding affixes to the end of the word, e.g. *receiv**er***, *wektor**yzacja*** 'vectorisation'. *Wektoryzacja* is created by the verbalisation of the noun *wektor* 'vector', which leads to *wektor-yzować* 'to vectorize'. Then, the verb *wektoryzować* is nominalised and the word *wektoryz-acja* is derived. The affixes that follow the word stem from which the new word is derived are called suffixes.

Other types of affixes cover infixes, circumfixes, transfixes and reduplication. Circumfixes and reduplication are not common in English and Polish and for this reason they will not be discussed here. Infixes may be encountered in quite a few Polish verbs in the perfective aspect, e.g. *da-wa-ć* 'to give' (from *dać* 'to give'), *obiec-yw-ać* 'to promise' (from *obiecać* 'to promise'). Polish has many transfixes, which are discontinuous affixes that occur in more than one place in the word, e.g. *u-łatw-i-ć* 'to make easy' (from *łatw-y* 'easy'), *na-słoneczn-i-ć* 'to expose to the sun's rays' (from słoneczn-y 'sunny'). Both *ułatwić* and *nasłonecznić* are perfective verbs and refer to actions that have been completed. Infixes and transfixes may be found only in verbal constructions, which do not occur in the Polish surveying termbase as they refer to activities and nouns rather than verbs are central in the termbase, e.g. instead of *centrować* or *wycentrować* 'to centre' the termbase includes *centrowanie* 'centering'.

Prefixes and suffixes comprise a highly diversified group in English and Polish. Plag (2003, p. 98) provides a semantic classification of English prefixes into four classes: quantifying prefixes, locative prefixes, temporal prefixes and negative prefixes. This classification scheme may be easily adapted for Polish, as many English prefixes have equivalents in Polish which are similar or identical in form. Plag's listing of prefixes is presented in Table 3-1. In this table I look only at prefixes and where they occur in the words. I will look at the whole internal structure of the words later on in this chapter. The items marked with asterisks in the table are non-standard prefixes in Polish and require more discussion, which will be provided in the text below the table. Empty spaces on either the English or Polish side of the table mean that no equivalents exist for a given prefix.

Table 3-1 Classification of prefixes

English prefix	example	Polish prefix	example	meaning of the prefix
quantifying prefixes				
uni-	unilateral	jedno-	jedno+częstotli-wościowy 'single-frequency'	one
bi- di-	bilateral disyllabic	dwu-	dwu+częstotli-wościowy 'double-frequency'	two
multi- poly-	multi-purpose polysyllabic	wielo-	wielo+językowy 'multilingual'	many
semi-	semi-conductor	pół-	pół+przewodnik 'semi-conductor'	half
micro-	microorgan-ism	mikro-	mikro+organizm 'microorganism'	small
macro-	macrostruc-ture	makro-	makro+struktura 'macrostructure'	large
hyper-	hyperlink	hiper-	hiper+nowoczesny 'hypermodern'	over
super-	superhuman	super-	super+market 'supermarket'	above, beyond
ultra-	ultramodern	ultra-	ultra+fiolet 'ultraviolet'	beyond, extremely
under-	underpay	niedo-	niedo+rozwój 'underdevelopment'	not suffi-ciently

sub-	subset	pod-	podzbiór 'subset'	under, below
vice-	vice-chancellor	wice- pro-	wice+prezydent 'vice-president' pro+dziekan 'vice-dean'	instead of, in place of
arch-	archbishop	arcy-	arcy+dzieło 'master piece'	chief
		przy-	przy+długi 'lengthy'	quite
		prze-	prze+piękny 'most beautiful'	most
		locative prefixes		
circum-	circumnavi-gate	o-	o+krążyć 'to circle'	around
endo-	endocentric			internal to
epi-	epicentral			on, over
inter-	international	między-	między+narodowy 'international'	between
intra-	intravenous	do-	do+żylny 'intravenous'	inside
para-	paralympics	para-	para+olimpiada 'paralympics'	along with
retro-	retrospection	retro-	retro+spekcja 'retrospection'	back-wards
under-	underground	pod-	pod+ziemny 'underground'	below
trans-	transatlantic	trans-	trans+atlantycki 'transatlantic'	across
		nad-	nad+morski 'seaside'	at
		za-	za+graniczny 'foreign'	behind
		przy-	przy+biblioteczny 'located by the library'	by

temporal prefixes				
ante- pre-	antenatal premature	przed-	przed+wojenny 'pre-war'	before
post-	post-war	po-	po+śmiertny 'posthumous'	after
neo-	neoclassical	neo- nowo-	neo+klasyczny 'neoclassical' nowo+żytny 'modern'	new
negative prefixes				
a(n)-	asymmetri-cal	a-	a+symetryczny 'asymetrical' nie+symetryczny 'not symetrical'	not
anti-	anti-abortion	anty-	anty+semita 'anti-Semite'	against
counter-	counteract	kontr- przeciw-	kontr+atak 'counter-attack' przeciw+działać 'counteract'	against
de-	deactivate	de-*[1] unie-*	de+aktywować 'deactivate' unie+szkodliwić 'deactivate'	to make (sth) not (e.g. working/h armful)
dis-	disapproval	dez-*	dez+organizacja 'disorganisation'	lack of (sth)
in- non- un-	inactive non-standard unrelated	nie-	nie+aktywny 'inactive' niestandardowy 'non-standard' niezwiązany 'unrelated'	not
		bez-	bez+względny 'ruthless'	without

The interesting fact about Polish locative prefixes which do not have English equivalents is that they are adjectival prefixes created from prepositional phrases, e.g. *nadmorski* 'seaside' is created from the PP (prepositional phrase) *nad morzem* 'over the sea', i.e. at the seaside, *zagraniczny* 'foreign' is created from the PP *za granicą* 'behind the border', i.e. abroad.

[1] * symbol indicates that a form is semantically and/or grammatically incorrect

The Polish equivalents of English negative prefixes *de-* and *dis-* are also noteworthy. As an equivalent of *deactivate* the form *deaktywować* is encountered in Polish but it is a very recent borrowing. The standard Polish equivalent is *unieszkodliwić*, which is an example of multiple affixation as the base word *szkodliw-y* 'harmful' is negated in *nie-szkodliw-y* 'harmless' and then the transfixation process takes place to form the verb *u-nieszkodliw-ić* 'to make sth not harmful', i.e. to deactivate. In *dezorganizacja* 'disorganisation' the prefix *dez- is encountered. Dez-* is not a standard Polish prefix but one that was created on the basis of the English prefix *dis-* and means 'lack of sth'. The case of *asymmetrical* is quite similar to the one of *disorganisation*, but in Polish both *asymetryczny* 'asymetrical' and *niesymetryczny* 'non-symetrical' are equally frequent. Furthermore, the English prefixes *in-*, *non-*, *un-* have a single equivalent in Polish, which is *nie* 'not'.

Suffixes are more numerous than prefixes. Szymanek (1988) proposes a classification of suffixes on the basis of their semantics. He provides a list of 25 cognitive categories (Object, Substance, Person, Number, Existence, Possession, Negation, Property, Colour, Shape, Dimension, Similarity, Sex, Space, Position, Movement, Path, Time, State, Process, Event, Action, Causation, Agent and Instrument), and tries to find the relationship between derivational categories and cognitive categories by establishing a list of cognition-based derivational categories valid for English and Polish. The categories are divided according to the syntactic class membership of the derived word (Szymanek, 1988, p. 112). Szymanek's classification of suffixes is far too elaborate for the purpose of this thesis and I will only classify them according to the syntactic category of the base and the derived word. Therefore, I will follow Plag (2003, p. 86) who divides suffixes into four groups:

- nominal suffixes, which are employed to derive nouns from verbs, adjectives and nouns, e.g. *-tion* as in *animation* (from the verb *to animate*), *-owanie* as in *centrowanie* 'centring' (from the verb *centrować* 'to centre');
- verbal suffixes, which derive verbs from other categories (mostly adjectives and nouns), e.g. *-ate* as in *originate* (from the noun *origin*), *-en* as in *broaden* (from the adjective *broad*), *-ify* as in *humidify* (from the adjective *humid*), *-ać* as in *chorować* 'to be ill' (from *chory* 'ill'), *wagarować* 'to play truant' (from *wagary* 'truancy');
- adjectival suffixes, which derive relational and qualitative adjectives, e.g. *-al* as in *azimuthal projection*, *-ed* as in *animated*

map, *-alny* as in *układ centralny* 'central system', *-owy* as in *drut inwarowy* 'invar rod';
- adverbial suffixes, which derive adverbs from adjectives and nouns, e.g. *-ly* as in *shortly*, *-wise* as in *lengthwise*, *-o* as in *szybko* 'quickly'.

Derivation and the termbase

Affixation processes are recorded in 236 out of 490 entries in the English surveying termbase and 268 out of 459 entries in the Polish surveying termbase. Prefixes are quite rare in comparison to suffixes and they comprise just around 10% of all affixes. The most common prefixes in the English surveying termbase are:

- *anti-,* e.g. *anti-spoofing;*
- *cross-,* e.g. *cross-correlation;*
- *inter-,* e.g. *interaction;*
- *multi-,* e.g. *multi-channel receiver;*
- *post-,* e.g. *post-processing;*
- *pseudo-,* e.g. *pseudorandom code;*
- *re-*, e.g. *reexpression;*
- *super-*, e.g. *superposition.*

The most frequent prefixes recorded in the Polish surveying termbase are:

- *bez-* 'without', e.g. *metoda bezwzględna wyznaczenia pozycji* 'absolute method for determining position', i.e. absoulte positioning method;
- *między-* 'inter-', e.g. *Międzynarodowy Ziemski System Odniesienia* 'International Terrestrial Reference Frame';
- *nie-* 'not', e.g. *stabilizacja nietrwała* 'non-permanent stabilisation', i.e. temporary stabilisation or marking;
- *jedno-* 'one', e.g. *odbiornik jednoczęstotliwościowy* 'one frequency receiver', i.e. single frequency receiver;
- *dwu-* 'two', e.g. *odbiornik dwuczęstotliwościowy* 'two frequency receiver', i.e. double frequency receiver;
- *wielo-* 'many', e.g. *wielotorowość sygnału* 'many paths of signal', i.e. multipath;
- *pseudo-*, e.g. *pseudoodległość* 'pseudorange';

- *meta-*, e.g. *metadane* 'metadata'.

Many terms from the Polish surveying termbase which indicate a native origin, are characterised by the presence of such prefixes as *od-*, e.g. *odbiornik, odbicie, odcięta, odległość, odniesienie, odwzorowanie*; *do-*, e.g. *dostosowanie*; *po-*, e.g. *powierzchnia.* The prefixes *od-*, *do-*, *po-* are in fact prepositions, meaning 'from', 'to' and 'past, on' respectively. The meaning of the derived words is non-compositional as it does not come from the meaning of the prefixes and verbal bases which comprise them. I will analyse how these words are made up below:

PREFIX	WORD	STEM
od-	'from'	*od-bior-nik* 'receiver'
brać	'to take'	od-bic-ie *'reflection'*
bić	'*to hit'*	od-cięt-a 'abscissa'
ciąć	'to cut'	od-leg-łość 'distance'
leżeć	'to lie'	od-nies-ienie 'reference'
nieść	'to carry'	od-wzor-ow-anie *'projection'*
wzorować	*'to'*	pattern' from the noun *wzór* 'pattern'
do-	'to'	do-stos-owanie *'adjustment'*
stosować	'to apply'	po- *'past, on'*
po-wierzch-nia	*'surface'*	wierzch '*top'*

It is worth noting that prefixes are mainly verbal. In most of the examples cited above, prefixes attach to imperfective verb stems to form perfective verbs, e.g. *od-ciąć* 'to cut', *od-nieść* 'to carry', from which nouns are derived, e.g. *odcięta, odniesienie.*

There are also words starting with *prze-*, e.g. *przesunięcie*, *przewaga*, *przewyższenie*; and *z-*, e.g. *zniekształcenie*, which may look like prefixed words but, in fact, *prze-* and *z-* are indicators of the perfective status of the verbs from which the nouns in question are derived.

WORD	STEM
przesun-ięcie	'shift'
przesun-ąć,	perfective of
sun-ąć	'to shift'
przewag-a	'advantage'
przeważ-yć,	perfective of
ważyć	'to weigh'
przewyższ-enie	'height difference'
przewyższ-yć,	perfective of
przewyższyć	'to outweigh' from
wyższy	'higher'
zniekształc-enie	'distortion'

zniekształc-ić,	perfective of
zniekształcić	'to distort' from
niekształtny	'distorted'

There are two types of suffix in the English and Polish termbases: nominal suffixes and adjectival suffixes. The English termbase contains such nominal suffixes as:

- *-ance*, e.g. *private conveyance system;*
- *-ence*, e.g. *interference;*
- *-er*, e.g. *transmitter*;
- *-ing*, e.g. *bearing*;
- *-ion*, e.g. *encryption*;
- *-ity*, e.g. *probability;*
- *-ment*, e.g. *measurement*;
- *-or*, e.g. *oscillator.*

Adjectival suffixes which occur in the English termbase include:

- *-able*, e.g. *visual variable;*
- *-al*, e.g. *accidental error;*
- *-ar*, e.g. *circular error probable;*
- *-ed*, e.g. *animated map;*
- *-ic*, e.g. *bathymetric map;*
- *-ing*, e.g. *engineering map;*
- *-ive*, e.g. *relative position.*

The most frequent Polish nominal suffixes are:

- *-acja*, e.g. *komparacja* 'comparison';
- *-anie*, e.g. *szacowanie nieruchomości* 'evaluation of immovables', i.e. real estate evaluation;
- *-enie*, e.g. *tyczenie* 'marking out', i.e. setting out;
- *-er*, e.g. *skaner* 'scanner';
- *-ica*, e.g. *węgielnica* 'a device used for measuring building corners', which are called *węgły* in Polish (sing. węgieł), 'optical square';
- *-nik*, e.g. *odbiornik* 'receiver';
- *-ość*, e.g. *nieruchomość* 'sth which is not mobile, immovable', i.e. real estate;

- *-owa*, e.g. *ogniskowa* 'related to the focus', i.e. focal length;
- *-ówka*, e.g. *krokówka* 'related to a step', i.e. measuring by means of steps.

The most common adjectival suffixes in the Polish surveying termbase are:

- *-czy* with variants, e.g. *łata miernicza* 'measuring rod', i.e. ranging rod;
- *-ły*, e.g. *stabilizacja trwała* 'permanent stabilisation';
- *-ny* with variants, e.g. *mapa topograficzna* 'topographic map', *niwelacja precyzyjna* 'precise levelling', *błąd systematyczny* 'systematic error', *mapa katastralna* 'cadastral map';
- *-owy*, e.g. *mapa kropkowa* 'dot map';
- *-ski*, e.g. *równik niebieski* 'celestial equator'.

Quite a few words in the termbase are multiply affixed, e.g. *anti-spoofing*, *cross-correlation*, *nie-ruchom-ość* 'immovable', i.e. real estate. Multiply affixed words raise the question of how they were derived. In order to understand this issue, the examples of *unregretful* and *nieruchomość* are scrutinised. The word *unregretful* may be analysed in two ways:

[un- [regret ful]]
[[un- regret] –ful]

The correct structure may be chosen by assuming that meaning is compositional. On this basis, *unregretful* must be analysed as 'not regretful' so *-ful* was the first affix to be attached to *regret* and then the meaning was negated by adding the prefix *un-*. The other analysis is not correct as **unregret* is an impossible construction in English.

Nieruchomość can be analysed in a similar way:

a. [nie- [ruchom ość]]
b. [[nie- ruchom] -ość]]

The word *ruchomość* is a noun in Polish meaning 'movable things'. Although the form ruchomość is morphologically correct as it is formed in the same way as other Polish abstract nouns by adding the ending –ość to the adjectival base, which in this case is ruchom-y 'movable', the word ruchomość is only used in context of nieruchomość to indicate the difference in the meaning of the two words. By contrast, the negative form *nieruchomość*, is a term in Polish civil law, meaning 'real-estate'. Thus, analysis a may be excluded on semantic grounds. Analysis *b* proves to be more relevant as it indicates that the noun *nieruchomość* 'non-movable', i.e. real-estate is derived from an adjective *nieruchomy* 'non-movable' to which the ending *–ość* was attached.

B. Conversion

Conversion is a type of derivation without any overt marking. It may also be referred to as zero-affixation or transposition (Plag, 2003, p. 12). Conversion belongs to non-concatenative word formation processes, i.e. such processes which do not depend on the creation of complex words by linking together bases and affixes. Conversion is a common process in English, where four types of conversion listed in (12) may be identified:

(12) a. noun to verb, e.g. *the book* and *to book*
b. verb to noun, e.g. *to drink* and *drink*
c. adjective to verb, e.g. *open* and *to open*
d. adjective to noun, e.g. *poor* and *the poor*

In Polish, conversion does not imply that the form remains the same because inflection leads to different endings for nouns and verbs. The most common type of conversion in Polish is the adjective to noun conversion. It is due to the fact that an adjective may play the role of the subject in the sentence as in (13).

(13) *Chorzy* powinni znajdować się pod stałą opieką lekarską.
'*The sick* should be under permanent medical care.'

However, other types of conversion listed in (12) may also be found in Polish and they are presented in (14):

(14) a. noun to verb, *e.g. poziom* 'the level' and *poziom-ować* 'to level';
b. verb to noun, *e.g. wykład-ać* 'to lecture' and *wykład* 'the lecture' or *wystaw-ić* 'to exhibit' and *wystaw-a* 'the exhibition';
c. adjective to verb, *e.g. błękitn-y* 'sky-blue' and *błękitn-ieć 'to become sky-blue'*;
d. adjective to noun, e.g. *styczny* 'tangent' and *styczna* 'tangent line'.

The other types of conversion which change the syntactic category of the derived words involve (Grzegorczykowa, Laskowski & Wróbel, 1998, p. 368):

(15) a. noun to adjective conversion, e.g. *człowiek* 'man' vs *człowieczy* 'human';
b. conversion of prepositional phrases to nouns, e.g. *bez sensu* 'without sense' and *bezsens* 'nonsense'.

In (15a), where *człowiek* is the stem, the consonant alteration (k : cz) may be observed. In fact, alterations are common in conversion. A full overview of different types of alterations may be found in section 3.1.1.

Conversion in Polish may be also encountered among some feminine names of professions and roles. As a rule, the feminine counterparts of the masculine occupations are created by adding the suffixes *-ka*, *-ini*, *-ica* to a lexical base, e.g.:

(16) a. *nauczyciel* 'teacher' - *nauczyciel-ka* 'female teacher';
b. *bóg* 'God' - *bog-ini* 'goddess';
c. *zakon-nik* 'monk' - *zakon-nica* 'nun'.

However, Sękowska (2002, p. 182) claims that there are some feminine names of professions which cannot be created in accordance with this rule. Instead, the noun maintains the masculine form but it gets a feminine interpretation, e.g. (pani 'Mrs') *redaktor* 'editor', *doktor* 'doctor', *inżynier* 'engineer', *magister* 'master' (Sękowska, 2002, p. 182). Thus, the noun is not specified for gender.

I followed Plag (2003) in recognising directionality as the main problem of conversion. Plag suggests four possible ways by which the direction of conversion may be established: by looking at the etymology of the words, by establishing which of the words is semantically more complex, by looking at the inflection and stress and by checking the frequency of occurrence. The last three criteria can be relatively easily challenged as they depend on the type of corpus. Checking which word came first in a reliable source, such as the *Oxford English Dictionary (OED),* provides some sort of evidence for establishing the direction of conversion. It is important to note, however, that the *OED* only documents what people managed to find on a particular word and whether one word was recorded as being used before another may be subject to chance.

A special type of conversion is proper name generalisation, in which proper nouns become common nouns. Proper name generalisation is a type of coinage as it leads to the creation of neologisms. It incorporates two types of processes:

- processes in which trade names for commercial products became general words, e.g. *aspirin, vaseline, xerox;*
- processes in which the name of a person or a place after some time is used as a general word (Yule, 2010, p. 54), e.g. *hoover* (initially *spangler*), comes from the name of its inventor, Mr Spangler, who sold his idea to Mr Hoover and the product name was changed to his name. Words created in this way are called eponyms. Examples of eponyms involve *sandwich*, which comes from the eighteenth-century Earl of Sandwich who first insisted on having his bread and meat together, and *gal*, the unit of acceleration, which comes from the Italian physicist, mathematician and astronomer Galileo Galilei.

Some eponyms are general language words, e.g. *sandwich*, while others, based on the names of those who invented things, are technical terms, e.g. *volt* (from the Italian Alessandro Volta), *fahrenheit* (from the German, Gabriel Fahrenheit) or *watt* (from the Scottish inventor, James Watt).

Conversion and the termbase

Conversion is a rather marginal word formation process in the termbases. Noun and adjective verbalisation do not occur in either of the termbases as nouns are the main focal point in terminological analysis. In the English termbase, verb nominalisation is exemplified by:

- *cold start* and *warm start*, being the start modes of a GPS receiver, which come from the verb 'start';
- *land use map*, where the noun 'use' meaning usage comes from the verb 'to use'.

The English termbases also includes examples of adjective to noun conversion, e.g. *tangent*. There are also a few entries in the English termbase for which the direction of conversion cannot be established by looking at semantic complexity. The origin of such terms as *level*, *plan* or *track* can be traced in the *OED*, which gives the first attested occurrence of words. According to this source, *track*, *level* and *plan* were initially nouns and substantially were used as verbs. On this basis, the direction of conversion may be established as noun to verb. In the Polish termbase, there are a few cases of adjective to noun conversion, e.g. *celowa* 'related to target', i.e. line of sight (from *celowa-ć* 'to target') and *styczna* 'tangent' (from *styka-ć 'to touch'*, with the consonant alteration k : cz). There are also a very few instances of verb to noun conversion, e.g. *odstęp* 'interval' (from *odstęp-ować* 'to depart from', i.e. to abandon).

C. Compounding

Compounding, along with derivation, is a highly productive word formation process. It relies on combining two or more lexical bases in order to create a new lexeme, e.g. *bench mark*, *gwiazdozbiór* 'star collection', i.e. constellation. Following ten Hacken (1994, p. 25), I will present the compound structure as $[X\ Y]_Z$, where Z is a compound and X and Y are its components.

When the structure of compounds is analysed, one may notice many issues in their construction such as headedness, the order of elements in compounds, linking elements, feature percolation or recursion (Bauer, 2009, p. 343). As the first issue, I will discuss headedness. Many of the compounds consist of a head and a modifier. Semantically, the head of a compound specifies the class of entities to which the meaning belongs (Katamba, 1993). Its identity can be discovered by hyponymy. For example, the compound *coordinate system* is a hyponym of *system*. Syntactically, the head is the dominant constituent of the construction, which means that the inflectional properties of the compound are inherited from the head element, e.g. *ground segment* receives plural inflection on the second member only: *ground segments*, not **grounds segment*. There are also compounds which do not have an obvious head, e.g. the English compound *girlfriend*. The order of elements cannot be changed in this compound and it is the position of the external inflection that indicates which element is treated as the head.

As the next issue, I will discuss the order of elements in compounds. English headed compounds are mainly right-headed, e.g. *base map*, *user segment*. Polish compounds may be one-word expressions, e.g. *gwiazdozbiór* 'star collection', i.e. constellation, or multi-word expressions, consisting of a noun and a relational adjective (N+RA), e.g. *punkt wezł-owy* 'point which is a node', i.e. nodal point, or two nouns, where the head is in the nominative case and the modifier in the genitive case (N+GN), e.g. *godło mapy* 'emblem of the map', i.e. map nomenclature. Polish compounds which are one-word expressions are right-headed, e.g. *gwiazdozbiór* with *zbiór 'collection'* being the head, while compounds which are multi-word expressions are left-headed with *punkt 'point'* and *godło* 'emblem' as heads.

The next issue to be discussed is the linking element. Compounds in many languages have an element that makes their structure explicit but does not affect their meaning. This element is called a linking element (ten Hacken, 1994, p. 29) or an interfix (Dirven & Verspoor, 2004, p. 71). The linking element may be similar to an inflectional ending, e.g. *n* in German *Sonnenschein* 'sunshine'. Linking elements do not play a significant role in English. The possessive marker can be analysed as a linking element, e.g. *children's film*. However, they are quite common in Polish compounds. Polish compound nouns (17) contain a linking vowel *-o-, -i-/-y-* or *-u-* and the linking vowels in Polish compound adjectives (18) are usually *-o-* or *-u-* (Szymanek, 2009, p. 466).

(17)

STEM 1	STEM 2	COMPOUND NOUN
gwiazd-a	*zbiór*	*gwiazd-**o**-zbiór*
'star'	'collection'	'constellation'
star-y	*druk*	*star-**o**-druk*
'old'	'print'	'antique book'
łam-a-ć	*strajk*	*łam-**i**-strajk*
'break'	'strike'	'strike-breaker'
dw-a	*głos*	*dw-**u**-głos*
'two'	'voice'	'dialogue'
czworo	*bok*	*czworobok*
'four'	'side'	'quadrilateral'

(18)

STEM 1	STEM 2	COMPOUND A
ciemn-y	*brąz-ow-y*	*ciemn-**o**-brąz-ow-y*
'dark'		'dark-brown'
dw-u	*maszt-ow-y*	*dw-**u**-maszt-ow-y*
'two'		'two-masted'

Czworobok 'quadrilateral' is a very interesting compound because it looks like a compound with a linking element between the two bases, but in fact it does not include a linking element but a collective numeral of *czworo* 'four'. Collective numerals in Polish occur with *pluralia tantum*, e.g. *czworo drzwi* 'four doors', nouns which indicate people of both sexes, e.g. *troje ludzi* 'tree people', nouns which name immature individuals, e.g. *pięcioro szczeniąt* 'five puppies' and with *oko* 'eye' and *ucho* 'ear' in the plural, e.g. *dwoje oczu* 'two eyes'. A similar structure may be found in the compound *mimośród* 'eccentricity', which does not have a linking element either. Its interpretation is based on the historical meaning of its components. It consists of two bases: *mimo* whose current meaning is 'in spite of, past' and whose historical meaning is 'close to, next to', and *śród* meaning 'środek', i.e. centre.

Another issue in the structure of the compounds is feature percolation, which means that the compound as a whole inherits features from the head (Plag, 2003, p. 136). This property is reflected in the inflection of compounds. If the English endocentric, nominal compound is pluralised, the plural marking occurs on the head, not on the modifier. Thus, the plural form for *park commissioner* is *park commissioners*. The form *parks commissioner* is also correct, but the plural interpretation is limited to the modifier and not inherited by the whole compound. In the case of *deep-fry* with its past form *deep-fried,* it may be noticed that the head bears the inflection of the entire compound. Gender percolation may be easily found in Polish compounds, e.g. *sygnał radiowy* (N+GN) 'radio signal' is neuter. The compound consists of two components: *sygnał* 'signal', which is a masculine head and *radiowy* 'radio', which is a *relational adjective* created from the word *radiowy* 'radio', being the neuter modifier. The compound inherits the gender from its head. The plural form of *sygnał satelitarny* is *sygnały satelitarne* 'satellite signals'. The denominal adjective *satelitarny* agrees in number and gender with the head it modifies.

Polish has quite a few exocentric compounds whose gender does not come from the gender of its constituents. In fact, a shift of the gender class can be recorded in such compounds (Szymanek, 2009, p. 469). For example, it can be a shift from feminine to neuter or masculine gender as in (19).

(19)

STEM 1	STEM 2	COMPOUND
wod-a 'water'(f)	*głow-a* 'head' (f)	*wod-o-głow-ie* 'hydrocephalus' (n)
płask-a 'flat'	*stop-a* 'foot' (f)	*płask-o-stop-ie* 'flat foot' (n)
czarn-a 'black'	*ziemi-a* 'Earth' (f)	*czarn-o-ziem* 'black Earth' (m)

The last property of compounds to be discussed here is recursion, which means that a compound can be an element in another compound

(Lieber, 2009a, p. 350). The rules of compound formation can create the same kind of structure repeatedly by stacking new words on an existing compound (Plag, 2003, p. 134). Recursion is very common in English and a number of examples can be found in the surveying termbase, e.g. *International Terrestrial Reference Frame, European Geostationary Navigation Overlay Service*. In fact, five-member compounds are the longest ones I could identify in the English termbase. The five-member compound can be analysed as in (20) using the bracketing structure.

(20) [European [Geostationary [Navigation [Overlay Service]]]]

The majority of multi-word expressions in Polish consist of two words, e.g. *łata niwelacyjna* (N+RA) 'levelling staff', *nieoznaczoność fazy* (N+GN) 'ambiguity of phase', i.e. ambiguity of phase cycles. Therefore, recursion is not a very common phenomenon among Polish compounds in the surveying termbase. There are only 39 cases of recursion in the Polish surveying termbase, most of which occur in compounds consisting of three elements, e.g. *laserowe pomiary satelitarne* (RA+N+RA) 'laser measurements based on satellites', i.e. laser satellite ranging, *ewidencja podatkowa nieruchomości* (N+RA+GN) 'register of tax on properties'. There are only two compounds consisting of four elements: *natężenie pola siły ciężkości* (N+GN+GN+GN) 'intensity of field by force of weight', i.e. gravitational field intensity', *Państwowy System Odniesień Przestrzennych* (RA+N+GN+RA) 'State System of Reference to Space', i.e. National Spatial Reference System. The termbase includes only one case of a five-member compound *Geodezyjna Ewidencja Sieci Uzbrojenia Terenu* (RA+N+GN+GN+GN) 'Geodetic Register of Network for Infrastructure of Land', i.e. Spatial Registration of Utility Infrastructure. The reason why recursion is not very frequent in the Polish surveying termbase is the frequent use of qualitative adjectives (QA) in multi-word units which qualify the formation as a syntactic combination, not as a compound, e.g. *Główny Geodeta Kraju* (QA+N+GN) 'Chief Surveyor of the Country', where *główny* 'chief, main' is a qualitative adjective.

Compounds are often considered as the interface between morphology and syntax (Scalise & Bisetto, 2009, p. 35), which is particularly true in the case of synthetic compounds. Synthetic compounds constitute a special group of compounds, which can be analysed either as compounds or as derivations e.g. *meat-eater, blue-eyed* (ten Hacken, 1994, p. 26). In both types of analysis, one of the constituents has further structure. In a compound analysis, they are:

[meat] [eat er]
[blue] [eye ed]

In a derivational analysis they are:

[[meat eat] er]
[[blue eye] ed]

Bloomfield (1933, p. 231) defends a derivational analysis claiming that forms like *long-tailed*, *red-bearded* and *blue-eyed* derive from phrases like *long tail*, *red beard* and *blue eye*, from which they differ by the presence of the suffix *-ed*. The compound *meat-eater,* although it looks like synthetic compound, differs from other compounds discussed here because the phrase *meat-eat* does not exist. One can find a phrase *eat meat* or the word *eater*. Therefore, Bloomfield classifies this compound as a syntactic compound and claims that *meat* and *eat* can only be compounded if *-er* is added at the same time.

Polish has many compounds which in line with Bloomfield's theory could be analysed as derivations. These are compound nouns and adjectives which include affixes. For example, *[[obco-kraj]owiec]* 'foreigner' consists of a noun phrase *obcy kraj* 'foreign country' and a suffix *-owiec* and *[[prosto-kąt]ny]* which consists of a noun phrase *prosty kąt* 'right angle' and suffix *-ny*. Such components of the compounds as *krajowiec* and *kątny* do not exist as words in Polish, which accounts for Bloomfield's derivational analysis as in the case of *long-tailed.*

There is another case where components of a potential compound do not correspond to words. Bloomfield (1933, p. 242) observed that such a situation occurs in the case of 'foreign-learned vocabulary'. In many languages, there are words which behave unusually in certain morphological processes. These items are loanwords with a 'learned' connotation and encompass affixes, words and elements like *electro,* that are somehow in between these two. Ten Hacken (1994, p. 27) classifies such elements as stems. Furthermore, he argues that if an item may occur both as first and last element of a word without difference in meaning, this item is a bound stem. For example, *radio**phony*** and ***phono**logy* are based on the same Greek stem *phono*, which can be used both in the initial and final position in the word.

Compound classification schemes

The classification of compounds is a topic widely discussed by Bloomfield (1933), Jespersen (1942), ten Hacken (1994), Sękowska (2002), Lieber (2009a), Plag (2003), Scalise and Bisetto (2009). Compounds may be classified according to various sets of criteria. The most obvious criterion that can be derived from the compound structure is the nature of the modifier. On the strength of this criterion compounds may be classified into:

- regular compounds in which the modifier is a word, e.g. *clock correction, łata niwelacyjna* 'levelling staff';
- phrasal compounds in which the first element is a phrase or even a sentence and the second element is a noun, e.g. *fly-by method, mapa do celów projektowych* 'map for design purposes'. The range of phrases that appear as non-head of a compound is not limited to lexicalised phrases (Wiese, 1996, p. 187). In fact, the modifier can be a phrase from a different language, e.g. the Polish term *metoda on-the-fly* includes the English phrase 'on the fly';
- neoclassical compounds, which are formed on Greek and Latin bound roots, e.g. *geoinformation, aerotriangulacja* 'aerotriangulation'. Neoclassical elements can combine with other neoclassical elements to form compounds, e.g. *geology* or with words, e.g. *geodata* (from *geo-* 'Earth' and *datum* '(thing) given'). Neoclassical elements can appear as left- or right-hand elements in a compound, e.g. *geo-*, *hydro-*, *photo-*, etc., or *-logy*, *-metry,* or in either position, e.g. *philo* which is the right-hand element in *anglophile* and the left-hand element in *philology*. Left-hand elements of neoclassical compounds usually end in the vowel *-o*, e.g. *kilometre* sometimes -*i*, e.g. *centimetre*. The status of *-o* or *-i* is not immediately obvious, as it may be treated as a part of a suffix attached to the left-hand element of the neoclassical compound, a prefix attached to the right-hand element of the neoclassical compound, or a linking element (Plag, 2003, p. 157). Neoclassical compounds are technically oriented and very frequent in terminology. Neoclassical compounding will be discussed as a part of the extensive process of neoclassical word formation in section D.

As the next issue, I will dicuss more advanced theories on classification schemata. Bloomfield believes that differences between languages are so prominent that there is no scheme of classification that would fit all languages (Bloomfield, 1933, p. 233). According to Scalise and Bisetto (2009, p. 37) there are two problems with existing classifications. One of them depends on the fact that labels, e.g. 'root compound', 'synthetic compound' are not applicable to all languages, and the other is caused by the large variety of criteria adopted for classification.

The analysis of different compound classification schemes discussed by Scalise and Bisetto allows for the conclusion that headedness is the central criterion in any classification system. This line of classification allows for the distinction between the following three types of compounds (ten Hacken, 1994: 38, following Bloomfield):

- endocentric compounds, which have a head constituent, e.g. *solar day* is a kind of day, *punkt kontrolny* 'control point' is a kind of point;
- exocentric compounds, which lack a head element, e.g. *loudmouth, pickpocket*. Exocentric compounds have a metonymic meaning, e.g. *loudmouth* is used to refer to a person who talks a lot especially in an offensive or stupid way. Exocentric compounds are sometimes called bahuvrihi compounds, which is a Sanskrit term literally meaning '(having) much rice' (Bloomfield, 1933);
- copulative compounds (sometimes called *dvandva compounds*), both elements of which play the role of the head, e.g. *singer-songwriter*, *kobieta-żołnierz* 'woman-soldier'.

If exocentric compounds are scrutinised, it may be noticed that they are only semantically exocentric, as other features of these compounds relate them to endocentric constructions. For example, the part of speech or the number of the whole exocentric compound is inherited from the right-hand member. Thus, the plural for *loudmouth* is *loudmouths* and the word is a noun, just like a *mouth*. Exocentric compounds do not have transparent meaning. If *skinhead* is taken as an example, it may be recognised that it does not denote the *skin on the head* but a young person, usually a man, who has very short hair or no hair, and belongs to a group of often violent people. Benczes reckons that such compounds are metaphor- and metonymy-based (2010, p. 219). Metaphor may be defined as analogy between two objects or ideas, conveyed by the use of one word instead of another, e.g. *space ship*, where *ship* was used to denote a vehicle that travels in space. Metonymy was defined by Lakoff and Johnson (1980, p. 35) as 'using one entity to refer to another that is related to it', e.g. *white-collars* are used to refer to people who work in offices as they usually wear white shirts.

Developing the classification of compounds on the basis of the headedness criterion further, Bloomfield (1933) subdivides endocentric compounds into subordinative constructions, in which one member belongs to the same form class as the entire construction, and coordinative constructions, in which more than one element belongs to the same form class as the construction. Subordinative constructions are determinative compounds, such as *blackbird* (the head of the compound – *bird* belongs to the same class of nouns as the whole compound), and compounds that correspond to coordinative constructions are copulative compounds, e.g. *bittersweet* (both elements belong to the same class as the compound, which is the adjective class).

Scalise and Bisetto (2009, p. 45) elaborate on the grammatical relations that are possible between the two constituents of a compound and divide them into subordination, attribution and coordination. Subordinate compounds may be defined as compounds whose two constituents are in a head-complement relation e.g. *bridleway, efemeryda pokładowa* 'ephemeris of a board', i.e. broadcast ephemeris.

Attributive compounds are those compounds in which one element stands in a relation of attribute or modifier to another (Lieber, 2009b, p. 88). Attributive compounds consist of different formations. These formations are made up of a noun-head that can be modified by an adjective' or by a noun or a verb, e.g. *civil engineering*, *kartodiagram liniowy* 'linear cartodiagram'.

Coordinate compounds include constituents that can be connected by the conjunction *and.* These compounds are characterised by two equal elements whose order may be reversed without changing the meaning of the whole compound, e.g. *general-statesman* and *statesman-general.*

Lieber (2009a, p. 359) develops this classification scheme of compounds further by adding a syntactic interpretation to the coordinate, subordinate and attributive compounds.

As for coordinate endocentric compounds, she argues that they can have either simultaneous interpretation, as in the case of N+N compounds (*producer-director*), V+V compounds (*stir-fry*), and A+A compounds (*deaf-mute*), or they can have a mixture reading, as in the case of A+A compounds such as *blue-green.* Coordinate exocentric compounds can also have different interpretations. One can distinguish between the N+N compounds, e.g. *parent-child* (relationship) and A+A compounds *English-French* (negotiations) that have a relationship interpretation, as they consist of constituents that are semantically similar in some way. Apart from them, there are also coordinate exocentric compounds with collective interpretation, which are represented by N+N coordinates, e.g. *father-*

daughter (dance) or compounds with disjunctive interpretation which, however, are very uncommon, e.g. V+V compound *pass-fail*. Ten Hacken (1994, p. 67) calls them copulative compounds and claims that they have a symmetric relation between their elements.

The category of subordinate compounds is even more complex, especially when it comes to endocentric compounds. Lieber (2009a, p. 361) provides a three-step classification of endocentric compounds. First, she divides them on the basis of the presence/absence of verb constituents into verb-containing and verbless. Then she distinguishes different types of verb-containing compounds, viz. synthetic compounds, $[V]_N$+N compounds, which either contain a constituent that is a verb or one that has undergone conversion from a verb to a noun, $[N+V]_V$ compounds, which are the result of back formation, and N+$[V]_N$ compounds, which include compounds whose first constituent is a noun and whose second constituent is also a noun, but formed by conversion from a verb. Finally, Lieber classifies all types of verb-containing compounds according to the orientation into: object-, subject- and adjunct-oriented. Lieber's classification of subordinate endocentric verb-containing compounds is presented in Table 3-2.

Table 3-2 Classification of subordinate endocentric verb-containing compounds after (Lieber, 2009a, p. 361)

	synthetic	$[V]_N$+N	$[N+V]_V$	N+$[V]_N$
object-oriented	truck driver	kick-ball	head-hunt	chimney sweep
subject-oriented	city employee	attack dog	machine-wash	sunrise
adjunct-oriented	home-made	skate park	spoon-feed	boat ride

Subordinate endocentric verbless compounds consist of simple nouns in the head position, which have a relational or processual interpretation and often require complementation e.g. *table leg* (leg of a table), *cookbook author* (author of a book).

Subordinate exocentric compounds are not very common in English and there are a few examples that include verbs as the first constituent and nouns as the second: *pickpocket*, *cutpurse*, *spoilsport*.

Attributive compounds are generally claimed to be the most productive compounds in English (Lieber, 2009a, p. 362). There are four types of attributive endocentric compounds: N+N compounds, e.g. *file cabinet*,

A+N compounds, e.g. *blackboard*, N+A compounds, e.g. *ice cold* and A+A compounds, e.g. *dark blue,* and two types of exocentric compounds: N+N compounds, e.g. *birdbrain* and A+N compounds, e.g. *redhead.* Exocentric compounds can be metonymic, e.g. the compound *birdbrain* does not actually mean a *brain* of a *bird* but is used to refer to a foolish person.

The grammatical relation between the two constituents of the compound is not a crucial criterion in the classification of surveying terms because this research focuses on the meaning of terms. For this reason, this classification scheme is briefly introduced here and is not applied generally in the analysis of compounds which were identified in the surveying termbases.

The next compound classification scheme to be discussed is salient in the analysis of terms. It was developed by Jespersen (1942) and it classifies compounds according to the syntactic category of the head into:

- nominal compounds, e.g. *eyepiece, legenda mapy* 'map legend';
- verbal compounds, e.g. *sleep-walk*, *lekceważyć* 'to weigh lightly', i.e. to disregard;
- adjectival compounds, e.g. *sugar free, prostokątny* 'with right angle', i.e. rectangular.

On many occasions it is not easy to differentiate between compounds and syntactic combinations. The classification of compounds provided by Sękowska (2002, p. 167) may be of some help in this matter. Sękowska distinguishes three types of compounds:

- compounds with a linking element *-o-* which may have a suffix, e.g. *cartogram*, *fazomierz* 'phase meter', *aerotriangulation, średniowieczny* 'middle-aged';
- compounds which do not have a linking element and whose components are written together, e.g. *bridleway*, *footpath.* In Polish, the first element of the compound is usually uninflected, e.g. *Wielkanoc* 'Easter' (cf. $\text{Wielkanocy}_{\text{GEN}}$) vs *wielkiej nocy* 'great night';
- compounds which consist of two or more elements written separately, e.g. *standard deviation*, *księga wieczysta* 'land and mortgage register'. The order of constituents in such compounds cannot be changed because the meaning of the whole compound changes, e.g. *wieczysta księga* means 'a perpetual book'.

The first two cases of compounding may be easily recognised. It is problematic, however, to distinguish between compounds consisting of at least two elements written separately and syntactic phrases. In English, if all elements of such a construction are nouns, e.g. *spring balance*, *monitor station* the constructions may be recognised as compounds. In Polish this issue gets complicated, because, apart from such examples as *godło mapy* 'map nomenclature', and *antenna odbiornika* 'receiver antenna', which are indeed compounds, other formations which consist of two nouns, e.g. *rower dziadka 'grandfather's bike*, but which should not be recognised as compounds, may be found. According to ten Hacken (1994, p. 82), compound instances of Saxon genitive may be distinguished from non-compound instances by observing the behaviour of determiners. If in the structure *Determiner Noun's Noun*, the determiner agrees with the second noun, e.g. *the children's film*, the structure is a compound. However, if the determiner agrees with the first noun, as in *these children's film*, the structure does not involve a compound.

If a construction contains an adjective, one has to take into account whether the adjective is relational or qualitative. Formations containing nouns and relational adjectives may be classified as compounds and those which include qualitative adjectives are typically syntactic combinations (ten Hacken, 1994, p. 90).

Relational adjectives occur only attributively and cannot be modified with 'very' or 'more' (Lieber, 2005, p. 414), so *focal length* means 'a length referring to a focus'. It cannot be very focal or more focal. A qualitative adjective, on the other hand, can usually occur both in attributive and predicative position, e.g. *automatic level* (the level is automatic) and can often be modified by *very* and *more* (e.g. *very automatic level*). Relational adjectives are most often derived from nouns thanks to suffixes, e.g. *spherical aberration* (from sphere), *cadastral data* (from cadastre). However, some denominal adjectives are subject to both a relational and a qualitative reading, e.g. *criminal lawyer* has a qualitative meaning as a lawyer who is criminal and a relational interpretation as a lawyer who specialises in criminal law. There are no compounds with adjectives that have this type of ambiguity in the surveying termbases. Lieber (2005) makes some generalisations on tendencies that the English suffixes have with regard to relational and qualitative meaning. She believes that the *-al* suffix favours a relational reading, although a qualitative reading is also possible, e.g. *vocational course* may be interpreted as a course which helps one to find one's vocation and as a professional course which teaches new skills. Suffixes such as *-ed, -esque, -ful, -ic, -ish, -ly, -ous, -some, -y* seem to produce almost exclusively qualitative readings (e.g. *animated map, dynamic map*), but *geodetic data* and *geographic information* have a relational meaning. Finally, *-ive, -ory, -ant* create mainly relational adjectives, e.g. *relative positioning*. On the basis of this analysis, it may be concluded that suffixes may work as indicators of the relational and qualitative status of adjectives but each individual example needs to be analysed separately to make the right judgment.

The other criterion which differentiates compounds from phrases is stress pattern. While phrases tend to be stressed on the last word, e.g. *field of viéw*, compounds have their left-hand member stressed, e.g. *tópográphic map*. There are many systematic exceptions to this rule. According to Plag (2003, p. 130), copulative compounds which are compounds including two components referring to the same entity, e.g. *geologist-astrónomer,* have rightward stress. Other exceptions to the rule are compounds which contain temporal or locative modifiers, e.g. a *sidereal mónth, the London éye*, or a causative modifier which can be paraphrased as 'made of', e.g. *invár wire*, or 'created by' as in *Mercator projéction.* Ten Hacken (1994, p. 37) considers the stress pattern a problematic criterion as one expression may have different variants of stress assignment. For example, *ice cream* may have a stress assigned in two ways: as *'ice 'cream*, in which it is a phrase, and as *'ice ,cream*, when it is considered a compound (Bloomfield, 1933).

Compounds also differ from phrases semantically as they express concepts which are more specialised than concepts described by phrases, e.g. *blackbird* is a particular species of a bird, while the phrase *black bird* denotes any bird of black colour (Bloomfield, 1933, p. 227).

Compounds and the termbase

In the English surveying termbase I identified 238 out of 490 entries which are compounds. There are also around 25 syntactic combinations which include compounds, e.g. *apparent solar day* includes the compound *solar day*, *Doppler Ranging Integrated on Satellite* includes the compound *Doppler Ranging*, *fundamental bench mark* includes *bench mark*, *ground-based augmentation system* includes *augmentation system*, etc. In the Polish termbase, 268 out of 459 entries were recognised as compounds. Additionally, there are 10 syntactic combinations which include compound components, e.g. *Państwowy Zasób Geodezyjny i Kartograficzny* 'State Resources for Geodetic and Cartographic Data', i.e. National Geodetic and Cartographic Resources, which includes the compound *Zasób Geodezyjny* 'Resource for Geodetic Data', i.e. Geodetic Resources.

All headed compounds in the English surveying termbase are nominal compounds, which corroborates the fact that noun compounds are the most common compounds in English. In the Polish termbase the majority of compounds are RA+N formations, e.g. *tachimetr diagramowy* 'diagram tacheometer', *siatka kartograficzna* 'cartographic grid'.

There are also quite a few compounds in the Polish termbase which consist of two nouns, one of which is in the genitive case (N+GN), e.g. *segment użytkowników* 'users' segment', *spłaszczenie elipsoidy* 'ellipsoid flattening', *elipsoida Krassowskiego* 'Krassowski's ellipsoid'. In English, similar expressions in the form of combinations with possessive marker may be found. They may be reflected by a Saxon genitive, e.g. *Tissot's indicatrix,* or not morphologically marked as in *Doppler effect.* Di Sciullo and Williams (1987) claim that constructions with proper names are compounds only if the proper names they include are famous names. The reason why *Tissot's indicatrix* or *Mercator projection* are recognised as compounds, and constructions such as *Adam's dog* or *Mary's scarf* are not, is that the former ones have been generally accepted in the language and have been lexicalised. Ten Hacken (1994, p. 82) suggests that the difference between compound instances of the Saxon genitive and non-compound instances can be observed in the behavior of modifiers as in (21).

(21) a. Tissot's indicatrix
b. Adam's brother
c. Adam's older brother

Both Lees (1960) and Marchand (1969) classify items like (21a) as compounds. They have rigid structures and modifiers are allowed neither in front of them nor in the middle of them. In contrast, (21b) which also includes a proper noun, allows the insertion of an adjective that modifies a noun as in (21c) and therefore cannot be classified as a compound.

Apart from N+N combinations with proper nouns and possessive case there are also many random combinations which consist of two generic nouns, one of which is possessive, e.g. *dragon's den, students' union.* Whether a given expression can be assigned to the compound category depends largely on context and can be judged on the basis of determiners, which is illustrated in (22).

(22) a. the dragon's underground den
b. these dragon's dens

Whereas combinations such as *dragon's den* and *students' union* are considered to be compounds, due to the fact that they have been lexicalised in the language, inserting an adjective as in (22a) means that the expression is no longer a compound but a syntactic phrase. (22b) shows that the determiner agrees with the compound as a whole, not with *dragon* and confirms that in this case *dragon's den* is indeed a compound.

Other features of compounds will now be considered. The majority of compounds in the English surveying termbase are endocentric, but a few cases of exocentric compounds were also recorded. These are listed below:

- *eyepiece* - a type of lens that is attached to a variety of optical devices such as telescopes;
- *offset* - a short distance measured perpendicularly from a main survey line, created from the phrase 'to set off';
- *plumb bob* - a weight with a pointed tip on the bottom, that is suspended from a string and used as a vertical reference line;
- *scatterplot* - a discrete set of data;
- *total station* - an electronic instrument used in modern surveying which incorporates a theodolite with circles and an EDM;
- *waypoint* - a permanently stored and named position in the GPS receiver's memory.

There are also endocentric compounds, which have undergone conversion by removal of their heads and they look as if they are exocentric compounds. These compounds include:

- *choke-ring* (from choke-ring antenna) - a particular form of omnidirectional antenna for use at high frequencies;
- *fly-by* (from fly-by method) - a method of showing spatial change, which depends on using a sequence of views of a static surface or volume in which the viewpoint of the observer changes gradually;
- *rapid static* (from rapid static method) is a type of shortening used in surveying jargon; a type of static technique which involves the master remaining at its primarily control point while the rover is moved to each subsidiary point in turn;
- *real time kinematic* (from real time kinematic technique) is a type of shortening used in surveying jargon; it is the DGPS procedure whereby carrier-phase corrections are transmitted in real time from a reference receiver to the user receiver.

The termbases also contain endocentric compounds, which have metaphorical and metonymic meaning. These compounds include:

- *dead reckoning* - a way of calculating the position of a ship or aircraft using only information about the direction and distance it has travelled from a known point;
- *cold start* - the start mode of a GPS receiver in which the receiver is able to start providing position updates without the assistance of any almanac information stored in its memory;
- *warm start* - the start mode of a GPS receiver when current position, clock offset and approximate GPS time are known;
- *dumpy level* - a surveyor's level having a short telescope fixed to a horizontally rotating table and a spirit level;
- *spirit level* - an instrument designed to indicate whether a surface is level or plumb;
- *wye level* (Y level) - a level with Y-shaped rests to support the telescope;
- *remote sensing* - the science of acquiring information about material objects, area, or phenomena, without coming into physical contact with the objects, or area, or phenomena under investigation;
- *stadia hairs* - a pair of additional horizontal hairs engraved on the diaphragm of a theodolite.

Phrasal compounds do occur in surveying terminology but are very rare, e.g. *fly-by method.* Neoclassical compounds, on the other hand, are very frequent in the two termbases and will be discussed along with other neoclassical formations in section D.

D. Neoclassical word formation

Neoclassical word formation is word formation in which lexemes of Latin or Greek origin are combined to form new combinations that are not attested in the original languages (Plag, 2003, p. 155). It has its roots in the borrowing of a large number of Greek words, which happened mainly in the 17th and 18th centuries. Morphologically complex words, such as *astrology*, were then reanalysed and their components were used to create other words which do not occur in Ancient Greek and which came into existence in the 19th century, e.g. *astrocyte*. The word refers to a star-shaped cell of the neurological tissue in the central nervous system and according to the *OED* it is attested from 1898.

Petropoulou and ten Hacken (2002) assume that these neoclassical elements are the basis for the description of neoclassical word formation and call them neoclassical formatives (NCFs). In their view, such complex words of Ancient or Greek origin as *morphology* should be reanalysed as [*morpho+logo+y*] instead of [*morpho+logy*]. They believe that NCFs have a bound status. NCFs are used to form new lexemes (neoclassical lexemes) but they are not lexemes themselves. They do not have a syntactic category, but only a feature which does not have a syntactic distribution. They need to undergo morphological processes to have a syntactic category assigned. The most common of such processes is suffixation, which results in attaching one of the suffixes to neoclassical lexemes: *-y*, *-ic*, *ous*, *-ist*, *-ism*, *-itis*, *-ia*, etc. Neoclassical lexemes are created mostly by combining two NCFs followed by a suffix, e.g. [*geo logo -y*]. There are also some NCFs which include both prefixes and suffixes, e.g. *polytheism* [*poly- theo -ism*].

Sometimes it is quite difficult to distinguish between neoclassical compounding and derivation. The difference between affixes and NCFs is that an affix cannot combine with another affix to make up a new word, e.g. (e.g. **dis-ic*), while an NCF can combine with another NCF, e.g. *geology*. There are also some neoclassical elements such as *pseudo-*, *macro-*, *micro-*, *multi-*, *di-*, *poly-* which some morphologists consider as NCFs (Szymanek, 2009, p. 472), while others regard them as affixes (Plag, 2003, p. 98). In fact, these elements have moved from the neoclassical vocabulary and have become general prefixes. Some terms in the surveying termbase are no longer considered as compounds although they consist of two NCFs, e.g. *geodesy*, where *geo* means 'Earth' and *desy,* derived from *daien,* means 'to divide'. This may be caused by the fact that the neoclassical borrowing is not fully reanalysed. The first element was reanalysed as an NCF as it occurs in many other words, e.g. *geology*, *geometry*, *geoscience* and it has a contemporary reading. The second part was not reanalysed as an NCF and it is not used in any other words and does not have current interpretation. On this basis, *geodesy* is considered to be NCF lexeme but cannot be recognised as a compound in line with the theory which states that the word can be seen as a compound only if its two elements are meaningful.

Apart from derivation, it is also possible to turn NCFs into lexemes by conversion. For example, the English compound *photograph* is formed from the NCF lexeme *photographo*, which is a compound of *photo* and *grapho*, by deleting the final *-o* (Petropoulou & ten Hacken, 2002).

Neoclassical formatives and the termbase

NCFs are quite common in the surveying termbases as they make up 16 neoclassical compounds out of 490 entries in the English termbase (6.72% of compounds are neoclassical compounds) and 21 neoclassical compounds out of 459 entries in the Polish termbase (7.25% of compounds are neoclassical compounds).

The most frequent NCFs occurring in the initial position in the compounds are:

- *carto-* meaning 'card', e.g. *cartography*, *cartogram;*
- *geo-* meaning 'Earth', e.g. *geology*, *geodata;*
- *iso-* meaning 'equal', e.g. *isoline*, *isotherm;*
- *ortho-* meaning 'straight', e.g. *orthophotograph;*
- *photo-* meaning 'light', e.g. *photogrammetry;*
- *topo-* meaning 'place', e.g. *topology*, *topography.*

The most common NCFs in the final position in the compounds are:

- *-meter*, also occurring as *-metry* meaning 'to measure', e.g. *tachometer*, *photogrammetry;*
- *-grapho* meaning 'to draw, write', e.g. *photography*, *topography;*
- *-logy* (from Gk. *-logia*) meaning 'science', e.g. *topology*, *geology;*

The surveying termbases contain both neoclassical compounds consisting of classical elements only, e.g. *altimeter, cartogram, cartography, centimetre,* and neoclassical compounds comprised of one classical and one non-classical element, e.g., *geomatics*, *accelerometer.*

There are also neoclassical compounds in the termbases which contain three elements of classical origin, e.g, *tri-gono-metry* (*tri* 'three' + *gon* 'angle' + *metro* 'to measure' + *y*), *photo-gram-metry* (*photos* 'light' + *gram* 'to draw, to write' + *metro* 'to measure' + *y*). Ten Hacken (1994) indicates that in principle the two structures shown in Figure 3-1 are possible:

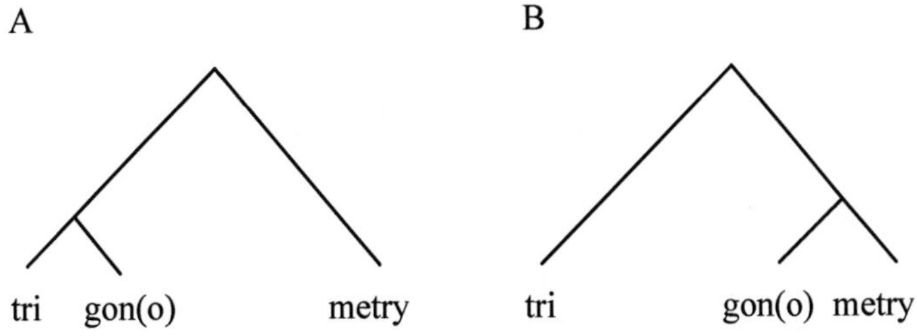

Figure 3-1 Structures of the neoclassical compound *trigonometry*

In structure A *trigon* is the intermediate node, and in structure B *gonometry* plays the nodal role. There are strong arguments to prefer structure A as *trigon* is a possible word meaning a triangle (cf. hexagon), and thus *trigono-metry* means measuring triangles. In contrast, *gonometry* is also possible but it is not semantically included in trigonometry. Even if I assumed that I am dealing not with *gonometry* but *goniometry* 'measurement of angles' which is a dictionary entry, the whole structure B would not make much sense as it would mean 'three measurement of angles'.

E. Shortening processes

According to Fischer (1998, p. 25) shortening processes involve a loss of material and lead to the creation of formations whose form and meaning are associated with those of base word(s). Shortening processes incorporate abbreviation, blending and clipping.

Abbreviations are amalgamations of parts of different words. Unlike affixation, they depend on losing material from the base word(s). Two basic types of abbreviation are initialisms and acronyms (Plag, 2003, p. 126). They are formed in a similar way by taking the initial letters of multi-word expressions, but they differ in pronunciation (Plag, 2003, p. 127). Initialisms are pronounced by saying each individual letter e.g. *GPS (Global Positioning System)*, while acronyms are pronounced regularly, just like words, e.g. *LIDAR (Light Detection and Ranging).* Acronyms in present-day English are orthoepically pronounced words (the stress is on the first syllable), typically made up of one or of two initial letters from at least two words. Some acronyms cannot be easily pronounced, therefore the initial letters of function words like *of*, *for*, *and*, *at* and *to* are omitted, e.g. *ASCII (American Standard Code for Information Interchange)*, *RUPP (Road Used as a Public Path),* or some additional letters are included, e.g.

MARV (Manoeuvrable Reentry Vehicle) instead of *MRV*. Acronyms tend to be written in upper case but there are also cases in which both upper and lower case letters are used to write the acronym, e.g. *SoHo (South of Houston Street)*. Such acronyms are numbered among those whose transparency is diminished. Apart from abbreviations composed of initial letters, there are also abbreviations which incorporate non-initial letters, e.g. *yd (yard), km (kilometre)*.

One may distinguish between terms which *have* shortened forms, e.g. *GPS (Global Positioning System)*, *CAD (Computer Aided Design),* and terms that *are* shortened forms as the longer forms are not used any more except to explain the name, e.g. *radar*, *lidar*, *sonar*. It is interesting that terms which are shortened forms are written in lower case. Many shortened forms are used more frequently than the full names, e.g. *DOP* with its variations (*HDOP, GDOP, VDOP* and *TDOP*), *GLONASS*, *DORIS*.

The second shortening process to be discussed is blending. Blending is the amalgamation of parts of different words. It depends on combining two (rarely three or more) words into one and deleting material from one or both of the source words. Plag (2003, p. 123) distinguishes two types of blends:

- blends which are existing compounds shortened to form a new word, e.g. *geoinformation* (from *geographic information*), *photointerpretation* (from *photographic interpretation*). In such blends, just like in compounds, the first element modifies the second element;
- proper blends, which denote entities that share phonological segments of the referents of both elements, e.g. *boatel* (from boat+hotel), which is both a boat and hotel.

Plag (2003, p. 132) considers that proper blends are formed according to the rule:

A B + C D → AD

This rule means that a new word is created by linking the first component of the first word with the final component of the second word, e.g.

Geo_graphy_ + _Mathe_***matics*** → ***Geomatics***
Sexag_enarian_ 'person sixty years old' + _de_***cimal*** → ***sexagesimal***

Plag (2003, p. 123) claims that the structure of blends is constrained by semantic, syntactic and phononological restrictions. The base words of the blend must be semantically related. Syntactically, the two words belong to the same grammatical category, mostly to nouns. Phonological conditions require that blends only combine syllable constituents and the size of blends is determined by the second element.

The next shortening process to be discussed is clipping. Clipping occurs by shortening the word by omission of one or more syllables. In most cases, the initial part of the base form is retained in a clipping, e.g.:

photo ← *photography*
auto ← *automobile*
lab ← *laboratorium* 'laboratory'
prof. ← *profesor* 'professor'

An interesting fact about clipping in Polish is that *prof.* ends in a non-final letter and therefore it is followed by a dot, whereas *mgr* clipped from *magister* 'master' ends in a final letter of the full form and does not receive a dot. This rule applies to all academic degrees in Poland.

The middle or final part of a base is maintained in clippings only if unwelcome associations or pre-existing identical or similar word forms make the retention of its initial part unpractical. For example, the clipping of *contact lens* is *lens*, not *contact*. Stress also plays an important role as the syllable which carries the primary stress is often kept, cf. *cred.* (credit). Moreover, there is a preference to retain the semantic nucleus of the base form, e.g. *bus* (omnibus) is more meaningful than *omni* and it is retained.

Shortening processes and the termbase

Shortening processes as a whole may be described as relatively frequent in the surveying termbase. Abbreviations, next to compounding and derivation, are the main word formation process. On the other hand, clipping and blending are much less frequent. Abbreviations are very common in English. Nearly 20% of terms included in the English surveying termbase are abbreviated. Initialisms are three times more frequent than acronyms. The English termbase includes 62 initialisms and 22 acronyms.

Abbreviations containing non-initial letters are also frequent. The great majority of them are shortened names of measure units, e.g.:

- *cm* (centimetre);

- *deg* (degree);
- *ft* (foot);
- *km* (kilometre);
- *mm* (millimetre);
- *yd* (yard).

Shortened forms include such items as:

- *digitisation* (map digitisation);
- *dynamic variable* (dynamic visual variable);
- *projection* (map projection);
- *transit* (transit theodolite);
- *visual variable* (static visual variable).

Finally, there are untypical abbreviations, which were created mainly on the basis of French names. These include:

- *SI* (International System of units) from French *Système International d'Unités*;
- *TAI* (International Atomic Time) from French *Temps Atomique International*.

Abbreviations are not as common in Polish as in English. I identified only 49 out of 459 entries in the Polish surveying termbase which have abbreviated forms. Moreover, many abbreviations in the Polish termbase have been borrowed from English. These include such initialisms as *AS, GPS, IGS, ETRF-89, ITRS, OFT, RTK, SA, SI, SLR, UTM, VLBI, WGS-84*. Acronyms borrowed from English incorporate the following items *CEP, DOP, EGNOS, EUREF, GLONASS, HDOP, POLREF (POLish REference Frame), PDOP, RINEX, TDOP, VDOP.*

Only 24 abbreviations are genuinely Polish. The majority of these abbreviations are shortened forms. There are only five Polish initialisms and six acronyms in the termbase. These initialisms are:

- *ASG-PL* (*Aktywna Sieć Geodezyjna-Polska* 'Active Geodetic Network-Poland');
- *CODGiK* (*Centralny Ośrodek Dokumentacji* Geodezyjnej i Kartograficznej 'Central Agency for Geodetic and Cartographic Records', i.e. National Mapping and Surveying Agency;
- *NMT* (*Numeryczny Model Terenu* 'Digital Terrain Model');

- *PKN* (Polski Komitet Normalizacji 'Polish Committee for Standardisation');
- *TBD* (*Topograficzna Baza Danych* 'Topographic Data Base'). Polish acronyms encompass the following examples:
- *GESUT* (*Geodezyjna Ewidencja Sieci Uzbrojenia Terenu* 'Geodetic Register of the Terrain Fitting Network', i.e. Spatial Registration of Utility Infrastructure;
- *GUGiK* (*Główny Urząd Geodezji i Kartografii* 'Main Surveying and Mapping Organisation');
- *SIG* (*System Informacji Geograficznej* 'Geographic Information System');
- *SIP* (*System Informacji Przestrzennej* 'Spatial Information System');
- *SIT* (*System Informacji o Terenie* 'Land Information System');
- *Dptr* for *dioptria* 'diopter' is an example of abbreviation with non-initial letters.

Shortened forms are very common in the Polish surveying termbase with such examples as:

- *biegun* 'pole' (*biegun geograficzny* 'geographic pole');
- *busola* 'compass' (*busola magnetyczna* 'magnetic compass');
- *efemeryda* 'ephemeris' (*efemeryda pokładowa* 'broadcast ephemeris');
- *legenda* 'legend' (*legenda mapy* 'map legend');
- *odwzorowanie* 'projection' (*odwzorowanie kartograficzne* 'cartographic projection');
- *poligon* 'polygon' (*poligon niwelacyjny* 'levelling polygon');
- *skala* 'scale' (*skala mapy* 'map scale').

Following Grzega and Schöner (2007) I consider such shortened forms as cases of ellipsis. Ellipsis is the deletion of a morpheme in an original composite form which served as a designation for the concept at issue. In the examples cited above, the determining part is left out.

Blending in the surveying termbases is the next issue to be discussed. Blends are not very common in surveying. The most frequent blends encountered in both termbases are in fact neoclassical compounds, whose modifiers have been shortened, e.g. *geoinformation (geographic information)*, *geodata (geographic data)*, *geosciences (geographic science)*, *fotointerpretacja* 'photointerpretation' (*interpretacja fotograficzna* 'photographic interpretation'), *fotopunkt* 'photopoint' (*punkt fotograficzny* 'photographic point'), *geoikonika* 'geoiconics' (*ikonika geograficzna* 'geographic iconics'), *geoinformatyka* 'geoinformatics' (*informatyka geograficzna* 'geographic informatics'), *geokodowanie* 'geocoding' (*kodowanie geograficzne* 'geographic coding'). Fisher (1998, p. 40) argues that, in cases like these, the actual blending is replaced by clipping of the modifier. When I look at Polish blends, e.g. *fotopunkt* or *geokodowanie*, I notice that their structure is the same as the structure of English blends and does not correspond to the structure of the Polish formations from which the blends derive (e.g. *geokodowanie* not **kodowaniegeo)*. Polish surveying blends, just like their English equivalents, are right-headed compounds, which are typical of English. On that basis, I may judge that blends in the Polish termbase have been actually borrowed from English.

The non-standard cases of blending encompass such examples as English *pixel*, *pseudolite*, *pictometry* and Polish *ortofotomapa* 'orthophotomap'. The term *pixel* was created by the amalgamation of the first elements of the words *picture* and *element*. The term consists neither of the full form of the head and shortened modifier as in the case of compounds, nor is it a proper blend. In fact, creation of this term involves two processes: clipping which will be discussed in the next section, and blending.

The term *pseudolite* is also a very uncommon blend as it is derived from a prefix *pseudo-* and the word *satellite.* The prefix is combined with the second element of the *satellite*, while the first part of this word (*satel*) is left out.

The term *pictometry* was created by clipping the neoclassical word *pictographic* to *picto* and blending the shortened element with the final neoclassical combining form *-metry* 'measure'. Thus, the neologism *pictometry* referring to the vertical, oblique and aerial photography automatically captured and georeferenced, is created.

The Polish term *ortofotomapa* 'orthophotomap' was created in three stages. First, the neoclassical initial combining form *orto-* 'ortho' (meaning 'straight') was combined with *foto* 'photo', which may be a neoclassical combining form, but in this case it is a clipped form of the word *fotografia* 'photograph'. *Ortofoto* 'orthophoto' means a geometrically corrected aerial picture which has a uniform scale. In the next step, the blend *ortofoto* 'ortophoto' was compounded with the word *mapa*, thus creating *ortofotomapa*, which is a map consisting of orthophotographs. The formation *ortofotomapa* is closer in its structure to the English terms than to the Polish ones, which indicates that the formation was in fact loan translation from English and its components were reanalysed and modified in Polish to correspond to Polish spelling and inflection.

Finally, I will look at clippings in surveying. The surveying termbases do not contain simple cases of clippings. Clippings are in fact not very frequent and, if they occur at all, the word formation process usually involves clipping of the word forms in the first stage and blending different parts of the words and entire words in the next stage, e.g. *geoikonika* 'geoiconics' (from *ikonika geograficzna* 'geographic iconics'). Moreover, it is not possible to find genuinely Polish examples of clipping and blending in the termbase.

F. Analogy-based processes

Analogy-based processes involve such cases as these in which new complex words are created without any existing word formation rules but on the basis of a single (or very few) model words (Plag, 2003, p. 37). For example, *air-sick* was coined on the basis of *sea-sick*, *cheeseburger* was coined on the basis of *hamburger*, etc. The process by which such words come into existence is called analogy, which may be presented as a proportional relation between words, as illustrated below:

a : b :: c : d
sea : sea-sick :: air : air-sick
ham : hamburger :: cheese : cheeseburger

The key issue of this analogy is that the relation between two items, e.g. *a* and *b* is the same as the relation between two other, corresponding items, e.g. *c* and *d*. Examples of analogy-based processes are *blog* which was created from *web log* and reanalysed as *we blog* by analogy to *re-blog* and *pictometry* from the termbase, which was created by analogy to *telemetry*.

web log : log :: re-blog : blog
tele : telemetry :: picture : pictometry

A type of analogy-based process is backformation, which is the name for deriving words by dropping what is thought to be an affix. An example of backformation may be the word *edit*, which was derived from *editor* by dropping the suffix *-or*. The word was created on the basis of a proportional analogy with such word pairs as *actor-act* (Plag, 2003, p. 37). Other examples of backformation are *televise*, created from *television* by analogy with the type illustrated by *revision-revise* and *donate*, created from *donation* by analogy with the type illustrated by *motivation-motivate*. Backformation is common neither in English nor Polish.

Backformation may be established by looking at historical records and checking which word came first. Etymological dictionaries or general dictionaries which include etymological information may be consulted to find the word origin.

Analogy-based processes and the termbase

Analogy-based processes are not very common in the surveying termbases. Apart from *pictometry*, which was derived by analogy with *telemetry*, I have not found any examples documenting the occurrence of these processes. I have not identified any cases of backformation in the English and Polish termbases, either, which is caused by the fact that backformation depends mainly on deriving verbs from nouns and in the termbases terms are either nouns or noun phrases.

G. Borrowing

Borrowing depends on adopting words from other languages. Witalisz (2002) and Zabawa (2008) distinguish between two types of borrowings:

- lexical borrowings, which are easily recognizable as they resemble or are exactly the same as foreign words, e.g. *weekend*, *pub*, *interfejs* 'interface', *komputer* 'computer';
- semantic calques, which borrow the meaning of a foreign word and attach this meaning to a native word or a word that was borrowed earlier but is already well established in a language. Semantic calques are not easily recognisable and are sometimes referred to as hidden borrowings (Lüdeling, Schmid & Kiokpasoglou, 2002). Examples of semantic calques are *szczyt* 'summit' (originally only 'peak of the mountain', currently it also refers to 'a very important meeting' or 'a prestigious position') or *zamek* 'fortress, castle' (originally a 'locking device', later 'a castle', nowadays also 'a zip') from German *Schloss*. Semantic calques occur more often if English and the native word are formally similar, as in the case of *aplikacja* and *application*, *promocja* and *promotion*. This formal similarity is caused by the fact that both the English and Polish words are earlier Latin borrowings (Witalisz, 2006). When Polish borrowed words from Latin in the 16th century, it took only one

> meaning of a polysemic word, while English or French took over more meanings. Therefore, English or French Latinisms tend to have broader meanings nowadays. The surveying termbases include a few semantic calques, e.g. *plan* 'plan' (initially 'drawing, sketch', later 'scheme of action') or *stopień* 'degree' (first meaning 'grade' and later 'measure of an angle').

Lexical borrowings cover adapted borrowings, whose pronunciation and spelling were adjusted to the standards of language into which they were borrowed, e.g. *altimetria satelitarna* 'satellite altimetry', and unadapted borrowings, which were taken over from a foreign language without any changes in their form, e.g. *Circular Error Probable.*

The category of hidden borrowings, apart from semantic calques, encompasses structural calques, which are direct translations of English phrases, e.g. *farma urody* 'beauty farm', *strefa zero* 'ground zero'. A special type of calque are hybrids, which consist of one native element and one element of foreign origin, e.g. *megaprzebój* 'mega hit', *hipernowoczesny* 'hyper modern', *ciucholand* 'shop with second-hand clothes', which includes the Polish component *ciuchy* meaning clothes in the colloquial language and the English element *land.*

Some borrowings are essential as they name foreign customs, e.g. *Halloween,* new technological advances, e.g. *DVD, komputer, espresso* (Witalisz, 2006). They name foreign concepts which, initially, did not occur in Polish and had to be borrowed. It would be ineffective to create neologisms in Polish and try to replace these borrowings as many of essential borrowings make up the jargon of a particular field and using them makes international communication easier. On the other hand there are also borrowings which are not essential from a linguistic point of view, as they have their equivalents in Polish. They play only an expressive role and are elements of popular culture, e.g. *czatować na Internecie z drinkiem w ręce* 'to chat on the Internet with a drink in your hand' (Witalisz, 2007). According to Grzega and Schöner (2007), the most important sources of borrowings in English are Latin (from the 6th century until today), Old Norse (8th to 11th century) and French (11th to 15th century). In present-day Polish the majority of borrowings come from English. Polish, just like other European languages, is heavily influenced by English. This influence is caused by language contact, and borrowings of this kind are defined by Bloomfield (1933, p. 458) as cultural borrowings. This type of language contact is mainly one-directional. English acts as a superior language from which the Polish language borrows not only vocabulary but also morphology (prefixes, e.g. *de-/dez-* as in *dezaktywować* 'deactivate', *re-* as in *refinansować* 'refinance', *eco-* as in *ecodevelopment,* etc.), syntax (e.g. the use of noun clusters instead of RA+N constructions, e.g. *auto szyby* 'car-windows' instead of *szyby samochodowe* 'car windows') and punctuation (use of English quotation marks instead of Polish ones). This superiority is the result of many factors. Zabawa (2008, p. 155) believes that English started to be considered not only useful but also fashionable. The development of modern technology, mainly the Internet, and the dominant position of the USA in the modern world were also important. In addition, contacts between English and Polish tightened after Poland joined the European Union.

The influence of Polish on English is also documented but it is limited to vocabulary with as few as 19 words of Polish origin noted in the Oxford English Dictionary, 2nd edition by Podhajecka (2002, p. 333-337). In comparison, there are about 2000 English lexical borrowings in everyday Polish (Witalisz, 2007), which counts 150 000 lexical units (Dubisz, 2002, p. 395).

Borrowings from English are relatively recent. Before 1989 Polish was influenced by Russian to a great extent. Dubisz (2002, p. 395) claims that Russian influenced Polish for six centuries. Therefore, many borrowings come from Russian as well. These borrowings are mainly reflected in syntactic calques, which are direct translations of phrases. Russian and Polish have very similar grammatical systems, and for this reason many of these calques are unnoticed.

Apart from English and Russian, permanent borrowings in Polish come from Latin, German and French. Czech also had a meaningful contribution to the Polish vocabulary as it influenced Polish for seven centuries, particularly in the initial stages of the language's development. The Italian language influence on Polish is less significant. It is reflected in the name of vegetables such as *kalafiar* from *cavalfiore* or *pomidor* from *pomodoro*, which were introduced to Poland in the 16th century by Bona Sforza, an Italian wife of the Polish king, Sigismund I the Old. There is also more recent influence of Italian on Polish, which is reflected in the names of drinks and dishes, e.g. *pizza*, *spaghetti* or *cappuccin*o. These words have become international words and may be found in other languages as well.

According to Sękowska (2002, p. 184) the internationalisation of the Polish language is a very intensive process. Borrowing of foreign lexemes is becoming more and more common. The actual derivational processes such as affixation are weakened by the influx of foreign elements and word formation patterns. In fact, borrowing is one of the most productive processes in Polish.

Borrowing and the termbase

Borrowing is a common word formation process in English surveying terminology, where 148 entries out of 490 were identified as borrowings. Neoclassical borrowings are often difficult to spot. Many Latin and Greek words found their way into English in the 17th or 18th century and became assimilated so that they are no longer regarded as borrowings, e.g. *vector* or *map*. Some neoclassical elements such as *pseudo-* or *micro-* were reanalysed in English and are currently used as affixes. Apart from Latin, the most common source of borrowing for English was French, e.g. *algorithm* comes from the French *algorithme* or *azimuth*, which comes from French *azimuth*. There are also many words which were borrowed from Latin through French, e.g. *chain*, which was borrowed from Old French *chaine*, while *chaine* was borrowed from Latin *catena*.

It is relatively easy to recognise native Polish formations in the termbase from borrowings as the spelling of the former is characterised by the occurrence of diagraphs (*ch*, *cz*, *dz*, *dż*, *dź*, *rz*, *sz*), e.g. *przewyższenie* 'height difference', *rozdzielczość* 'resolution' and diacritics (*ę*, *ą*, *ć*, *ń*, *ó*, *ś*, *ź* and *ż*), e.g. *błąd* 'error', *wskaźnik* 'indicator', *żabka* 'foot plate', *zdjęcie* 'photo'.

There are 163 borrowings in the Polish surveying termbase. Many of these borrowings have been borrowed from Latin or Greek and their forms remained the same as in the source languages, e.g. *antena*, *gleba*, *globus*, *legenda*, *libella*, *limbus*, *orbita*. The termbase also includes quite a few learned borrowings, whose origin is hard to establish as many modern languages have nearly identical words, e.g. Polish *almanach*, English *almanac*, German *Almanach*, French *almanac*, Spanish *almanaque*. Borrowings from English are very common in Polish, e.g. *skaning laserowy* 'laser scanning', *Globalny System Pozycyjny* 'Global Positioning System'. Unadapted borrowings from English constitute a very large group of borrowings in Polish with as many as 32 examples in 459 entries, which come mainly from the field of Global Positioning System, which is one of the most recent subfields of surveying. Polish users try to adapt these borrowings by creating formations which consist of the Polish translation of the English term which is followed by the English abbreviation of this term, e.g. *Międzynarodowa Służba Geodezyjna IGS* 'International Geodetic Service IGS'.

3.1.3 Multi-word units

This section examines how multi-word units (MWUs) are created in English and Polish. It also provides an overview of existing typologies of multi-word expressions, which leads to the development of a classification scheme of multi-word units that serves the purpose of this research. Finally, it discusses the relation between compounds and multi-word units.

Surveying terms compiled in the termbases are of two types. One may distinguish between single-word units, e.g. *postprocessing*, *trigonometry*, *geodezja* 'geodesy', *barwa* 'hue' and multi-word units, e.g. *road used as a public path*, *plumb bob*, *odczyt w przód* 'reading in the front', i.e. foresight. Multi-word expressions (MWEs) cover a wide range of word combination categories from the most transparent ones, such as phrases, to the less apparent ones, such as idioms and collocations. Some of these multi-word expressions existed in general language and have been adopted as terms in a particular field, e.g. *road used as a public path*, *green lane* while others have been specifically created to name the concept, e.g. *Global Navigation Satellite System*, *metoda najmniejszych kwadratów* 'method of least squares', i.e. least-squares method (Görög, 2006, p. 218). The former may be recognised by the fact that they have a different meaning in general language from that in the given domain. They may be also used in other fields outside the given domain with different meanings, e.g. *base station* is used not only in GPS but also in the area of wireless computer networking to refer to a radio receiver/transmitter that serves as the hub of the local wireless network. The latter, in contrast, are specific to a given domain and do not have any transparent meaning in the general lexicon, e.g. *height above ellipsoid* is a GPS term and does not name a concept in any other field.

Multi-word expressions may be created as the outcomes of numerous word formation processes, which may not be recognisable if the overall structure of the combination is analysed, but which become apparent when such a combination is splitted into components. For example, *aerial triangulation* involves such processes as compounding (*aerial triangulation* is an endocentric compound), borrowing of its constituents and derivation (*triangle*, *triang-ulate*, *triangulat-ion*), *Geographic Information System (GIS)* illustrates such processes as derivation, compounding and abbreviation.

Atkins and Rundell (2008, p. 166-167) claim that, although there is a large body of work on the classification of MWEs, it is not possible to establish boundaries between different types of multi-word items as the boundaries between them are very fluid. In the following I will consult existing typologies of multi-word expressions, e.g. by Cowie (1998),

Burger (1998), Bauer (1983) and Sag et al. (2002) to develop a kind of typology for MWEs. I will then consider forms occurring in the termbases I have compiled and try to categorise them according to this classification scheme.

Granger and Paquot (2008, p. 27) discuss two major approaches to the study of multi-word units: the phraseological approach and the distributional or frequency-based approach. The phraseological approach identifies multi-word units on the basis of linguistic criteria and is very common in lexicology and lexicography. This approach is reflected in a few typologies of phraseological units, e.g. by Cowie (1998), Mel'čuk (1998) and Burger (1998). Cowie's (1998) typology was a starting point for later classification schemes and for this reason I will discuss it here. Cowie divides word combinations into composites, which function at or below the sentence level, and formulae, which function as independent utterances. Composites are further subdivided into restricted collocations, figurative idioms and pure idioms. These three elements along with free combinations form a phraseological continuum, which has the most variable and transparent elements on one side and the most opaque and fixed elements on the other side. Restricted collocations feature restricted collocability and the specialised meaning of one of the elements, e.g. *perform a task, heavy rain.* Figurative idioms also preserve a literal interpretation apart from figurative meaning, e.g. *make a U-turn*. Pure idioms are semantically non-compositional, e.g. *kick the bucket*.

The category of formulae encompasses 'sentence-like' units, which work as sayings, catchphrases and conversational formulae (Cowie, 1998). Formulae are later subdivided into routine formulae, which play speech-act functions, e.g. *good morning, good luck* and speech formulae, which organise messages and indicate the attitude of a speaker or a writer, e.g. *do you know what I mean?*

Cowie's classification was further developed by Burger (1998) who added a category of structural phraseological units including word combinations that establish grammatical relations, e.g. *as well as*. Granger and Paquot (2008, p. 43) combine Burger's typology, which is the most complete one, with a distributional approach to multi-word units, which is corpus-driven or frequency-based. This resulted in a new typology of multi-word units that gives a full account of possible word combinations, which do not all fit predefined linguistic categories. This new typology assigns phraseological units to one of three categories: referential phrasemes, textual phrasemes and communicative phrasemes. Referential phrasemes convey a content message and refer to objects, phenomena or real-life facts. They include lexical collocations, idioms, irreversible bi-

and trinomials, similes, compounds, grammatical collocations and phrasal verbs. Textual phrasemes structure and organise the content of a text or discourse and cover complex prepositions and conjunctions; linking adverbials and textual sentence stems. Communicative phrasemes are used to express feelings or beliefs or to address interlocutors, either to focus their attention, include them as discourse participants or influence them. They include speech act formulae in the form of greetings, compliments, invitations, etc., commonplaces (sentences expressing tautologies, truisms and sayings, e.g. *enough is enough*), proverbs and slogans. Due to the fact that textual and communicative phrasemes do not convey any content messages and only organise information or express certain attitudes, they cannot make terms. Therefore, I will only discuss subcategories of referential phrasemes which can function as terms. Referential phrasemes include the following classes:

- lexical collocations, which are usage-determined relations between two lexemes in a specific syntactic pattern. They consist of the 'base' of the collocation and a 'collocator' which is semantically dependent on the base, e.g. *heavy rain, closely linked, true north, random error, księga wieczysta 'land and mortgage register';*
- idioms, which are phrasemes constructed around the verbal nucleus. Idioms feature semantic non-compositionality, which can result from metaphorical process, and lack of flexibility along with marked syntax. Examples of idioms are *spill the beans, to bark up the wrong tree*;
- irreversible bi- and trinomials, which are fixed sequences of two or three words that belong to the same grammatical category and are linked by the conjunction 'and' or 'or', e.g. *bed and breakfast, left, right and centre, Definitive Map and Statement, ewidencja gruntów i budynków* 'land and building register*';*
- similes, which are sequences of words that work as stereotyped comparisons, e.g. *as old as the hills, to swear like a trooper;*
- compounds, which are made up of two or more elements which have independent status outside these word combinations. They can be written together, separately or with a hyphen. They feature high degrees of inflexibility and carry meanings as a whole, e.g. *goldfish, suitcase, eye-piece, bench mark*;
- grammatical collocations, which are restricted combinations of a lexical and a grammatical word. They typically consist of verb/noun/adjective + preposition, e.g. *depend on*, *bored with*, *contribution to*;

- phrasal verbs, which are combinations of verbs and adverbial particles, e.g. *blow up, make out.*

The most common types of phrasemes that occur in surveying termbases are compounds, lexical collocations and irreversible bi- and trinomials. A few cases of phrasal verbs may also be also recorded. Idioms, similes and grammatical collocations do not occur in the termbases.

Sag et al. (2002) classifies multi-word expressions in a slightly different way. They distinguish between lexicalised phrases and institutionalised phrases. Lexicalised phrases have at least partially idiosyncratic syntax or semantics, or contain words which do not occur in isolation. Lexicalised phrases cover:

- proper idioms, e.g. *kick the bucket*;
- decomposable idioms, e.g. *spill the beans*;
- compound nominals, which include terminological MWEs, e.g. *accelerometer, centrum fazowe anteny* 'antenna phase centre';
- proper names, e.g. *Los Angeles, Royal Institution of Chartered Surveyors*;
- verb-particle constructions, e.g. *set out*;
- light verb constructions, e.g. *make a mistake, take a measurement.*

The difference between proper idioms and decomposable idioms lies in the notion of 'semantic compositionality' introduced by Frege (1892), which is a means of describing how the overall sense of a given form is related to its parts. Decomposable idioms, such as *spill the beans*, can be analysed as composed of *spill* in a "reveal" sense and *the beans* in a "secret(s)" sense. The overall compositional meaning is "reveal the secret(s)". Such analysis is not possible for proper idioms such as *shoot the breeze.*

Institutionalised phrases are not usually taken as elementary lexical units, which means that they are not taken as lexicalised forms, and for this reason they do not belong to the lexicon (Agirre et al., 2006). They follow only general rules of syntax, where the word meanings combine compositionally but cannot always be substituted by synonyms. They are often conventionalised and they take only one of the possible readings available (e.g. *traffic light* means 'stop light' and not 'turning light').

When the two classification schemes of multi-word expressions are scrutinised, some degree of similarity and overlap may be noticed, as both typologies have idioms, and compounds and phrasal verbs from Granger and Paquot's typology (2008, p. 27) correspond to verb-particle constructions from Bauer's and Sag's typologies.In addition, lexical collocations and irreversible bi and trinomials actually belong to institutionalised phrases. Combining these two classification schemes and applying them to terminology only, leads to the classification of MWEs presented in Figure 3-2.

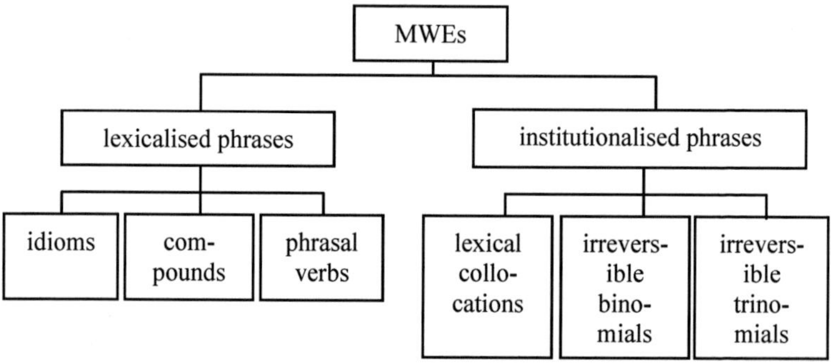

Figure 3-2 Typology of multi-word expressions.

At this point it is worth noting that due to the orthography of compounds it is often difficult to classify them as simple terms (written as single words) and complex terms (multi-word units). Plag et al. (2007, p. 95) distinguish between closed (solid) compounds, whose components are written together as in *backsight*, hyphenated compounds, whose components are linked with a hyphen, as in *cross-staff, fly-by, eye-piece,* and open compounds, whose elements are written separately as in *clock error.* There is no agreement in phraseology as to whether compounds should be considered as MWUs or not. The traditional view on phraseology excludes compounds from phraseology altogether (Barkema, 1996), or only keeps those units that meet well-defined criteria, such as stress, meaning, etc. Other views exclude solid compounds but include open and hyphenated compounds (Gläser, 1998). It is important to emphasise that, from the linguistic point of view, compounds are not multi-word units, but words as they are labels for single concepts. When a compound, e.g. *map projection* is compared to an idiom, e.g. *kick the bucket*, it may be noticed that the former one designates one concept, while the latter one refers to two separate concepts. I will not consider compounds as MWEs in my research.

MWEs are frequent in the surveying termbases. There are in total 490 entries in the English surveying termbase, 111 of which are MWUs. Compounds are common components of MWEs as 25 MWEs include compounds as their components. The Polish termbase includes 459 entries with 68 MWUs. There are 10 MWUs which include compound components. The most frequent MWUs in the two termbases are proper names, e.g. *British Cartographic Society*, *European Space Agency* and syntactic combinations which include a qualitative adjective and a noun, e.g. *accidental error*, *mapa zasadnicza* 'base map'. It is important to note that compounds are only orthographically MWUs. Semantically they are as simple single-word terms as they refer to one specific concept.

3.2 Terminological processes

This section discusses how English and Polish surveying terms are named. Grzega and Schöner (2007) specify three major processes of name giving: word formation, borrowing and semantic change. Following ten Hacken (2010a), I look at these processes from the onomasiological perspective and the semasiological perspective. While the onomasiological perspective starts from concepts and looks for their names, the semasiological perspective starts from words and asks for their meaning (Grzega & Schöner, 2007, p. 7). Just as everyone uses their language both as a speaker and as a listener, the two perspectives have to complement each other in linguistic description. The onomasiological perspective is more the perspective of a speaker who is looking for the name for the concept or idea he/she has in mind and the semasiological perspective is the perspective of a listener who is looking for the meaning of a word he/she has heard.

The onomasiological approach is more widely discussed in morphology and semantics than the semasiological approach. It is due to the fact that linguists working in the field of generative linguistics, which is concerned with forms only and therefore semasiologically-oriented, do not call themselves semasiologists. On the other hand, linguists working within the onomasiological approach describe themselves by saying how their perspective differs from the semasiological approach.

The term *semasiology* is often confused with *semantics* (Schmitter, 2008, p. 575). It is due to the fact that the term *semasiology*, which was invented in Germany and adopted in English-speaking countries, first indicated the meaning of the word form. It was gradually replaced by *semantics*. The year 1894 marks the point when *semasiology* started to be used with reference to a specific semantic perspective, called the *semasiological perspective* or *semasiological approach*. This perspective starts from forms and looks for their meaning.

Onomasiological theories of word formation have been elaborated by Miloš Dokulil, Ján Horecký, Pavol Štekauer, Bogdan Szymanek and an overview is given in Grzega (2009). I will not investigate all these theories as it is not the main subject of this thesis. I have selected Štekauer's model of word-forming or word-finding as a good, and well-elaborated, example to show the subsequent stages of the term-naming act.

Štekauer's cognitive onomasiological theory was inspired by Dokulil's onomasiological structure and by Horecký's multi-level model of linguistic sign. Its central idea is that word formation is a naming act performed by a speaker. Each naming act is a response to a naming need of a speech community (Štekauer, 2006, p. 35). A naming process consists of five levels:

- the conceptual level, where the concept is analysed and categorised as 'SUBSTANCE, ACTION, QUALITY and CONCOMITANT CIRCUMSTANCES (for example, Place, Time, Manner, etc)';
- the semantic level, where the semantic markers or components are structured;
- the onomasiological level, where one of the semantic components is chosen as the semantic base and the other as the onomasiological mark of this base. The semantic base is a class of objects to which a concept is assigned, and the semantic mark is a set of properties that differentiate this object from other objects in the same class. For example, the onomasiological basis of *car seller* is *-er* and its onomasiological mark is *car sell;*
- the onomatological level, where the morphemes are chosen;
- the phonological level, where the forms are combined.

The first three levels in this model of word-finding are purely cognitive and they prepare the speaker for making a decision on the onomasiological level of which word formation process to use to name the concept. The speaker can either juggle with existing forms and select one of the morphological word formation processes, or may use foreign words to name a given concept. I have already discussed word formation processes including morphological word formation and non-morphological word formation (borrowing) in some detail in section 3.1, therefore here I will discuss only a third process of name giving; semantic change.

Semantic change is a process in which no formally new creation occurs, but an already existing form is extended in use (Grzega & Schöner, 2007, p. 41). It exploits the polysemy of words (Blank, 2001). Semantic change uses various mechanisms, which have been categorised and discussed by Bloomfield (1933), Stern (1931), Ullmann (1962), Blank (1999) and Grzega and Schöner (2007). On the basis of these sources, I developed the following list of the most frequent types of semantic change:

- metaphor: a change of meaning based on similarity between concepts, e.g. *sole* (from Latin *solea*) first referred to the underneath of the foot, and later started to be used for the flatfish on account of the similarity of the shape (Cowie, 1998, p. 30);
- metonymy: change of meaning based on contiguity between concepts, e.g. *horn* 'animal horn' → 'musical instrument'. Sometimes a proper name is used to name a concept related to the name-bearer, e.g. *hoover* 'vacuum cleaner '[for which the company Hoover was an important producer] (Grzega & Schöner, 2007, p. 42). Metonymy may be difficult to distinguish from metaphor if only the definition provided by Grzega and Schöner (2007, p. 42) is taken into account, as the border between these two is not clearly specified. Therefore, I will rather use Cowie's (2009, p. 32-33) definition which delineates a metaphor as a "figure of speech that consists in using the name of one thing for the name of something else with which it is associated", e.g *caterpillar* for *track, shuttle* for *train,* and metonymy as a "figure of speech that consists in using the name of one thing for the name of something else with which it is connected in some respect", e.g. *sail* for *ship*, *crown* for *monarch;*
- synecdoche: change of meaning based on the 'part-of' relationship between concepts, e.g. *beam* 'log (of a tree)' ← 'tree (OE. beam)' (Grzega & Schöner, 2007, p. 42);
- specialisation or narrowing of meaning: change of meaning based on superordinate-subordinate relation between the old and new meanings, i.e. a term for a concept on a superordinate level is used to denote a concept on a subordinate level, e.g. *meat* 'any type of food' → 'the flesh of animals as opposed to the flesh of fish' (Yule, 2010, p. 233);
- generalisation or broadening/widening of meaning: change of meaning based on subordinate-superordinate relation between the old and new meanings; i.e. a term for a concept on a subordinate level is used to denote a concept on a superordinate level, e.g. *holy day* 'religious feast'→ *holiday* 'general break from work' (Yule, 2010, p. 233);

- cohyponymic transfer: horizontal shift in a taxonomy, e.g ModE *fir* → G. *Föhre* 'pine tree' (Grzega & Schöner, 2007, p. 42);
- antiphrasis: change of meaning based on a contrast between source and target concept, e.g. ModE. slang *perfect lady* → 'prostitute' (Grzega & Schöner, 2007, p. 42);
- antonomy, change of meaning based on a polar contrast between the source and the target concept that can be grouped on a kind of scale, e.g. *bad* in the slang sense of 'good' (Grzega & Schöner, 2007, p. 43).

There is a type of semantic change, called conceptual recategorisation, which is an onomasiological process. As a result of conceptual recategorisation a referent or a set of referents is assigned to another category, and what follows consequently it receives its designations, e.g. a community with 10,000 inhabitants (and which has a Cathedral) is a *city* in Britain, but a *town* in the US (Grzega & Schöner, 2007, p. 42).

Semantic change is a frequent process in term naming in surveying. I found a number of examples of metaphors, metonyms, generalisation and specialisation of meaning in the English and Polish termbases. They are listed below.

Metaphors:

- *pikieta* 'picket', i.e. picket point (a visible landmark, the position of which is determined with the use of a total station) comes from the French word *piquet* 'a painted stake, post or peg driven into the ground; used for various purposes. The first attested use of this term in surveying in English dates back to 1702 (*OED*). The general interpretation of this word, which is applied to 'people acting in a body or singly and who are stationed by a trades union or the like, to watch people going to work during a strike or in non-union workshops, and to endeavour to dissuade them' comes from 1867 (*OED*) and is a metaphor of *pikieta* in the picket point sense. *Pikieta* is also an example of meaning generalisation;
- *żabka* 'little frog', i.e. foot plate, denoting a heavy metal solid with a triangle-shaped base and two perpendicular pins on which a staff is placed, is a metaphor created because of its similarity to the shape of a small frog.

Metonyms:

- *pion* 'the perpendicular', i.e. plumb bob, denoting an auxiliary device employed for setting up straight lines and planes in a vertical position and for projecting points onto surfaces located on a height different from that of a given point.This is a metonym of *pion* 'the perpendicular', where the relation between the two words relies on the role of the tool;
- *szpilka geodezyjna* 'geodetic pin', i.e. arrow, which in Polish is associated with a tailor's pin (a small, thin piece of metal with a point at one end, especially used for temporarily holding pieces of cloth together), as, in surveying, it is used to mark each point where the measuring tape was put in the field. The term in Polish owes its name to the role it plays, while in English the name *arrow* was ascribed to this concept, taking into consideration these features of appearance which a geodetic arrow shares with an average arrow, i.e. a long thin stick with a sharp point at one end and often feathers at the other (a geodetic arrow has a red band instead of feathers).

Generalisation of meaning:

- *bench mark,* was first used in 1824 in surveying as a 'mark cut in some durable material, such as a rock, wall, gate-pillar, face of a building, etc., to indicate the starting, closing or any suitable, intermediate point in a line of levels for the determination of altitudes over the face of a country'. The word was transferred and acquired a figurative sense, meaning a level of quality which can be used as a standard when comparing other things. The first attested use of the word in this sense dates back to 1884 (*OED*);
- *channel*, originally denoting an old form of a canal. The meaning of the word was first generalised to 'connection for transfer' and then specialised to indicate a 'band of frequencies' in surveying;
- *godło* 'emblem', i.e. map nomenclature, appeared in the Polish lexicon in the 13th century and was used to refer only to national emblems or coats of arms (Boryś, 2005), currently it is used in the context of maps when speaking of their nomenclature, i.e. a numerical, or alphanumeric code that identifies a map unequivocally and defines its location in relation to other maps;
- *centr* 'centre', originally meaning the central point, it acquired a specialised meaning in surveying denoting an element on a sign

that indicates a correct geodetic point, the location of which is determined by X, Y coordinates resulting from network levelling;

- *legend*, initially meaning what is read (12th century), the story of the life of a saint (c. 1375), later on 'a story, history, account' (c. 1385), then a 'writing, inscription, or motto; chiefly in numismatics' (c. 1611), acquired its cartographic reading as the 'written explanatory matter accompanying an illustration, map' only in 1903 (*OED*);
- *chain*, originally denoting a 'connected series of links passing through each other, or otherwise jointed together, so as to move on each other more or less freely, and thus form a strong but flexible ligament or string'. This sense of the word dates back to c.1300. Its meaning developed over centuries and also acquired many figurative interpretations, e.g. c.1374. Chaucer used the word to refer to a binding or restricting force which prevents freedom of action, in 1397 it was used to denote a 'personal ornament in the form of a chain, worn a round the neck, in 1696 it started to be used as a 'series of individual acts, facts, events, or the like', in 1791 as a 'continuous linear series of material objects', and its surveying meaning goes back to 1610, when the word was used to refer to a 'measuring line, formed of one hundred iron rods called links joined together by eyes at their ends'. At first, chains of various lengths were used or proposed, but that described by the English mathematician, Edmund Gunter (1581-1626) in 1624 is the one now adopted; it measures 66 feet or 4 poles, divided into 100 links (*OED*).

Specialisation of meaning:

- *catenary*, which comes from the Latin word *catena* meaning chain, fetter, narrowed its meaning and it currently denotes the curve of an idealised hanging chain;
- *chart*, which was borrowed in 1571 from the Old French word *charte*, meaning map, card, developed its meaning and in 1696 it started to denote a sea-chart, which is a map used by navigators.

Blank (1999) tried to elaborate a list of motives for lexical change, which was revised and completed by Grzega and Schöner (2007, p. 24-30). I use this list to extract forces which cause semantic change. These forces involve:

- changes in the referent, when the designation of the concept remains the same, e.g *pen*, which is still used to denote a writing device but no longer one made of feathers (Grzega & Schöner, 2007, p. 24). This process is called substitution and it reflects a semasiological approach to name giving. However, the change in the referent may also involve introducing a new designation for a concept which is not completely new as it is based on already existing concepts, e.g. *brunch* as a mixture of a warm and cold meal between breakfast and lunch (Grzega & Schöner, 2007, p. 24). This process is onomasiological and is not a case of semantic change but of word formation;
- changes in the categorisation of the world which are reflected in the organisation of the concepts or the relevance of the referents, e.g. the word *girl* was originally used to denote a child of unknown origin and later on any unmarried woman. Its current meaning 'female human, teenaged or younger' results from a changing view on childhood and adolescence (Grzega & Schöner, 2007, p. 24);
- onomasiological fuzziness, which is a difficulty in classifying the concept or attributing the right word to the concept and leads to designations being mixed up, e.g. *arm* in the common usage in English means 'either of the two long parts of the upper body which are fixed to the shoulders and have the hands at the end' and its translation into Polish refers to 'the second part of the upper body between the elbow and the shoulder' (following Blank, 2001, p. 21);
- dominance of the prototype, which is a fuzzy difference between a superordinate and a subordinate term due to the prevailing role of the prototypical member of a category in the real world. It results in trademarks becoming general terms, e.g. *Kleenex*, specialisation or narrowing of meaning, e.g. *corn*, whose use was restricted from a general term denoting cereal to a term that refers to the type of cereal that is the most prominent in a region, such as oats in Scotland or wheat in England. It is also possible that a designation of the prototype serves as a basis for the designation of concepts, which are on the same hierarchical level, e.g. *apple*, which is a prototypical European fruit, served as a basis to create other fruit and vegetable names, e.g. pine apple. The last case scenario exemplifies the word formation process based on the use of a prototype (Grzega & Schöner, 2007, p. 30).

When I consider examples cited above, such as *chain, legend, chart*, I notice that, at some point in their etymological history, presented in the *OED*, their meaning became specialised and they acquired surveying interpretations. For example, *legend* was borrowed from French into English in the 12th century in the 'what is read' sense. The word's meaning became specialised in the 14th century and was used to refer to the biographies of saints or similar characters. In the 17th century, the word obtained an interpretation in numismatics denoting an inscription impressed upon a coin or medal. The cartographic interpretation as the written explanatory matter accompanying an illustration or a map appeared only in 1903.

As it may be noticed in this etymological analysis, the word *legend* was first used as a general language word when it entered the English lexicon permanently. Later, it started to acquire new, specialised senses, which were based on existing readings of the word, e.g. *legend* in the numismatic sense meant writing, inscription on a medal. This meaning was based on the ecclesiastical interpretation of legend as a piece of writing about saints.

The same regularity may be noticed in other word-formation processes, e.g. derivation or compounding. The term *pseudorange* was created by adding the prefix *pseudo-* to the noun *range*. Both elements were established in the English lexicon. *Pseudo-* comes from Greek but was reanalysed in English meaning 'false', and *range* dates back to the 13th century, initially meaning 'line, tow'. The sense 'scope, extent' dates back to the 1660s.

The term *chainman* was derived in the compounding process and first attested in English in 1824 in the surveying sense. The word structure is based on that of *fireman*, which appeared in the late 14th century to mean a 'tender of fire'. The sense 'person hired to put out (rather than tend) fires' comes from 1714 (*OED*). Hence, the word *fireman* first had to enter the lexicon permanently to be the basis for the creation of similar words denoting professions, e.g. *policeman, chainman.*

All the above examples are cases of lexicalisation, which is the process whereby a lexical item formed by a word-formation process is stored permanently in the lexicon shared by a community (Plag, 2003: 91, Cowie, 2009: 16). Lexicalisation is thus an expected step in word-formation. The meaning of the item which is permanently stored in the lexicon can be shifted or specialised, which is also called lexicalisation (Huddleston & Pullum, 2005, p. 288).

Lexicalisation usually occurs a couple of times in term naming. If the word is borrowed from another language and used with its original meaning, lexicalisation occurs once in the target language. The item enters the lexicon with the same form and meaning, e.g. *gleba* 'soil' (from Latin *gleba*), or its spelling is adapted to the spelling conventions of the target language, e.g. *zenith* from French *zénith.* However, lexicalisation also occurs at the level of the source language. The meaning of *gleba* was broadened as it used to mean a lump of soil, while the meaning of *zénith* was specialised as it initially meant 'road, path' (*OED*). If the expression is a native formation, it may be used either for the new, specialised meanings, which only involves specialisation of the meaning and its lexicalisation, e.g. *chart* or it may be created from the existing vocabulary in the word formation process and have its meaning specialised, e.g. *żabka* 'little frog', i.e. foot plate. In the latter case, lexicalisation occurs twice: after word formation, as *żabka* is a diminutive form of *żaba* 'frog' derived by suffixation, and after specialisation of its meaning.

3.3 Summary

In this chapter, I have analysed English and Polish surveying terms by looking at how they are formed and how they become named. Despite the fact that English and Polish differ greatly, as the former belongs to the Germanic languages, and the latter to the Slavic languages, there are quite a few tendencies in word formation which are common to the two languages. The most frequent word formation processes in English surveying terminology are derivation, compounding and abbreviation, while in Polish they are derivation, compounding and borrowing. In present-day Polish the majority of borrowings come from English, which is well-reflected in the surveying termbase with as many as 32 unadapted borrowings out of a total of 459 termbase entries. Polish and English surveying terms follow the same route of name giving, which occurs through lexicalisation with word formation, borrowing and semantic change being the intermediate steps in term naming.

CHAPTER FOUR

ANALYSIS OF CONCEPTS

This chapter concentrates on the meaning of terms. First, various theories of meaning and their features are examined (4.1) in order to establish the theoretical background for the semantic analysis of surveying concepts. In section (4.2) I discuss how ontologies can benefit knowledge organisation and the representation of concept systems and semantic relations between concepts. I then move to the presentation of conceptual mismatches resulting in lexical gaps which were found in the surveying termbases in English and Polish (4.3). The next section describes translation strategies for dealing with conceptual mismatches in general (4.4). Section (4.5) classifies translation problems encountered due to the occurrence of conceptual mismatches in surveying into classes with the same type of solution. The last section (4.6) provides a summary of the whole chapter.

4.1 Theories of meaning

'Meaning' is a vague term. In general, it refers to a variety of different relations between the world, language and speakers. There are many definitions of meaning which represent different views of its nature. Quine (1961, p. 47) points out that, without a satisfactory explanation of the notion of meaning, linguists working in semantics do not know what they are talking about. A definition of meaning is correct and complete only if it contains information on how meaning is realised.

There are three main approaches to meaning realisation:

- the referential approach, according to which meaning is a link between linguistic expression and objects in the real world (meanings as objects in the world);
- the communicative approach, which argues that meaning is constructed in communication by using words (meanings as uses);
- the mentalist approach, which claims that meanings are in the head (meanings as objects in the mind).

I use the description of theories of meaning as given by Riemer (2010) as a starting point for this discussion. I first elaborate on them in sections (4.1.1), (4.1.2) and (4.1.3), respectively. Then, I move to semantic relations in (4.1.4) and componential analysis in (4.1.5).

4.1.1 Referential theory

The referential theory of meaning is concerned with the relation between linguistic expressions and the world. The main component of the meaning of a linguistic expression is a referent or denotation (Riemer, 2010, p. 25). Thus, the meaning is a reference to facts and objects in the world. Filip (2008) points out that such a theory of meaning makes no psychological claims about the speaker's mind, mental objects, concepts or thoughts. It approaches natural language expressions as referring to things which are external to the concepts in people's minds.

The referential approach has its origins in the philosophy of language, logic and mathematics. The major works within this theory are those by Frege (1892), Russell (1905), Tarski (1933), Strawson (1950) and Montague (1970). Frege, Russell and Tarski were primarily interested in mathematical logic. Frege made a distinction between sense and reference and contributed to the compositionality principle (Szabó, 2007, section 1.5.4). Russell elaborated compositionality in the context of mathematics by developing what is now known as Russell's Paradox, which was used by Frege in his work on logic (Irvine, 2009, paragraph 4). Tarski developed a theory of truth for formalised languages by determining the criteria which the definition of a true sentence should meet (Hodges, 2010, paragraph 1). Strawson made his name with the article "On Referring" (1950), in which he criticised Russell's Theory of Description by pointing out that terms that refer to non-existent objects have a meaning. He is also famous for integrating the study of metaphysics with linguistic philosophy (Snowdon, 2009, section 2). Montague developed an approach to natural language semantics known as Montague Grammar, which states that semantically natural languages, like English, and formal logic should be treated in the same way (Pietroski, 2009, section 8).

Referential theories look at how words, expressions and sentences refer. Reference in general is defined as a relation that obtains between expressions and what speakers use expressions to talk about (Reimer, 2009, paragraph 1). It is questionable whether all expressions refer, but there are certainly several types of expressions which are of the referring sort. These expressions include:

- proper nouns, which refer to particular objects or individuals, e.g. *George W. Bush* refers to a particular man, *Barcelona* refers to a particular city;
- natural kind terms, which refer to kinds of things found in nature, which are things studied by scientists, e.g. *tiger*, *gold*, H_2O;
- indexicals, which are expressions where the reference depends on the context in which they are used; context here incorporates speaker, hearer, time and place, e.g. *I*, *here*, *there*, *he*, *she*, *now*, and *me.* They are also called deictic expressions or deictics (Riemer, 2010, p. 98);
- definite descriptions, which pick out a single individual uniquely, or which appear to do so, e.g. *the brightest student in the group, the first dog in space, the winner of the 2010 Nobel Prize for Literature.*

Frege (1892) argued that an expression's reference is not the only part of its meaning. He proposed a distinction between sense and reference as two different aspects of meaning. Thus, the reference [Bedeutung] is the thing that an expression refers to, e.g. the reference of *Barack Obama* is the man Barack Obama, while the sense [Sinn] is the way an expression picks out its reference. Frege calls it the expression's 'mode of presentation'. The sense of the expression is given by one or more definite descriptions. Hence, the sense of *Barack Obama* is given by the following definite descriptions: *the 44th President of the USA, the first black President in the history of the USA*.

Frege applied the distinction between sense and reference to solve the puzzle of identity statements (Riemer, 2010, p. 90), like those in (23).

(23) a. The Morning Star is the Evening Star.
b. The Morning Star is the Morning Star.

The Morning Star and *the Evening Star* have the same reference, which is the planet Venus. They have different senses, however, as the sense of *the Morning Star* is 'the last star to disappear in the morning' and the sense of *the Evening Star* is 'the first star to appear at night'. Frege shows that (23a) is different from (23b); (23a) is true and informative, while (23b) is tautological and trivial. Using identity statements, Frege proves that the expression's meaning consists not only of its reference, but involves a second part, which is called the sense. It is the sense which makes sentences (23a) and (23b) differ in meaning.

Senses and references have a number of features which allow us to distinguish between them. Senses are not mental and cannot be considered as psychological phenomena in anyone's mind. They are public and external to the mind.

Senses are not the same as ideas. Frege makes a clear distinction between senses and ideas, which he perceives as the 'internal images' a person associates with expressions. For example, the idea of the Queen of England is the mental image the person has of the Queen of England. Different people may have different mental images of the same idea but they can grasp the same sense.

Sense determines reference, which means that the reference of an expression depends on the sense of the expression. The sense of the expression contains a set of definite descriptions which means that the expression refers to one specific thing, person or place only, e.g. *Barack Obama* refers to the one specific person called Barack Obama. It may happen that two or more different people have the same name. Such proper names have different references and senses, however. They are a special case of homonymy. All meaningful expressions have a sense, but not all of them have a reference, e.g. *Santa Claus* has a sense ('the magical fat man who brings gifts to the homes of good children on the evening or night of Christmas Eve') but it does not have a reference since there is no such thing as Santa Claus.

Sense and reference are two aspects of meaning. The other two aspects of meaning which occur and exist together are denotation and connotation. Riemer (2010, p. 19) defines denotation of an expression as the entire class of objects, situations, etc. to which the expression correctly refers. Denotation is often used interchangeably with reference. It may be established by consulting a dictionary and checking the definition of the word. For example, the denotative meaning of *Hollywood* is 'area of Los Angeles, known as the centre of the US film industry'. Connotation, on the other hand, refers to the personal aspect of meaning, the emotional association that the word arouses (Kreidler, 1998, p. 45). Thus, *Hollywood* connotates such things as glitz, glamour, celebrity and dreams of stardom.

Meaning may be described fully by making reference to three principal terms: language, the world and the human mind. Ogden and Richards (1949) devised the 'semiotic triangle' to symbolise these three aspects of meaning. The semiotic triangle may be used in the referential and mentalist approaches. In the referential approach, it illustrates the relations between referent, symbol and thought (Figure 4-1), while in the mentalist approach it concerns the relations between object, word and concept.

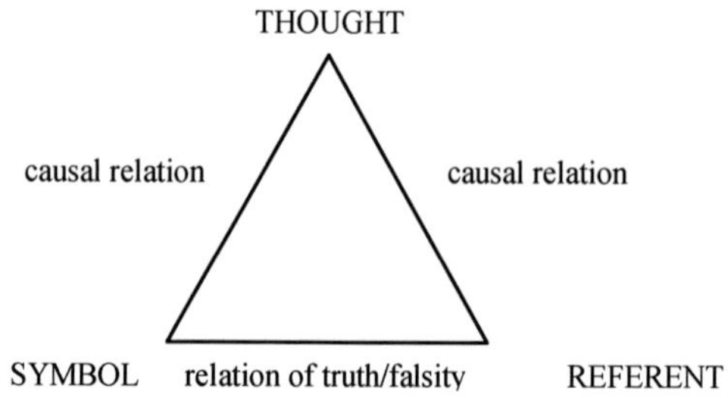

Figure 4-1 Ogden's and Richards' (1949) semiotic triangle

At the top of the triangle, there is thought, which reflects the fact that language comes from human beings. The bottom left corner of the triangle is the symbol, which is the token selected to express the speaker's meaning. In the case of spoken language, the symbols are speech sounds and in the case of written language, the symbols are characters. The bottom right corner of the triangle is the referent or the things, events and situations in the world to which the expression refers (Riemer, 2010, p. 14). The thought has causal relations with both the symbol and the referent. The causal relation between the thought and the symbol is justified by the fact that human minds use language to create linguistic expressions. The causal relation between the thought and the referent comes from the fact that when people use language, they intend their words to have a certain referent. On the symbol-referent side, there is no causal relation as there is no direct dependency between a string of sounds or signs and any particular referent. This feature is called arbitrariness.

The idea of arbitrariness was developed by Saussure (1916/1969, p. 66) who claimed that the arbitrary nature of the sign was the first principle of language. In his view, the linguistic sign is the combination of a signifier and a signified (Figure 4-2).

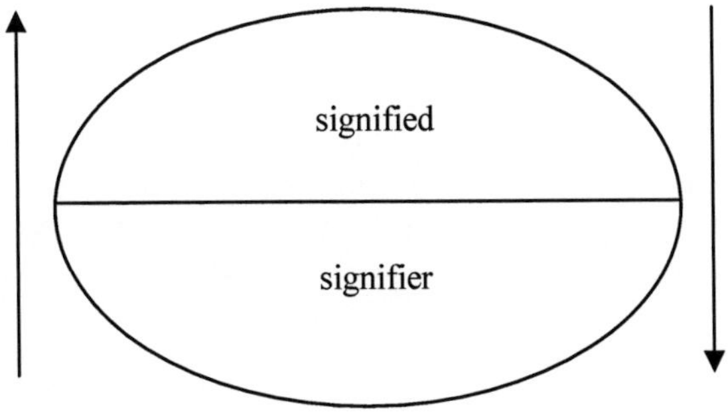

Figure 4-2 Saussure's model of the sign

The signifier refers to the sound image, which may be described as the psychological imprint of the sound or the impression it makes. The signified is the concept. The union of the two components of the linguistic sign is very close, as one part will instantly evoke the other, e.g. the sound image of tree in a given language will automatically evoke the concept TREE. The bond between the signified and signifier is arbitrary, as there is no natural, intrinsic, or logical relation between a particular sound image and concept. Therefore, there are different words, in different languages, for the same thing.

4.1.2 Communicative theory

The communicative approach to meaning states that a word meaning consists simply of the way it is used. This theory is also called the use theory of meaning (Riemer, 2010, p. 36). It was developed by behaviourist psychologists such as Skinner (1957), and linguists such as Bloomfield (1933). A slightly different form of the use theory was advanced by Wittgenstein (1953). Behaviourist theorists reject the notion that words have hidden, unobservable properties called meanings. They believe it is unscientific to use meanings in explanations since they are inherently unobservable. In their view, the only objective and scientific way to explain language is to use those features of language which are observable. These are the particular sequences of words and expressions which occur in actual examples of language use.

Bloomfield (1933, p. 139) reckons that the only meaning a linguistic form possesses is 'the situation in which the speaker utters it and the response which it calls forth in the hearer'. This may be illustrated with a very simple example using the word *sorry*. Its meanings may be described by the situations where a speaker apologises and the hearer accepts the apology and acts accordingly (e.g. by letting the incident pass without accusing the speaker of being rude, or by themselves saying 'sorry').

The main objection against the communicative theory of meaning lies in the great variety of sentences which make up the linguistic behaviour of an individual. There are some conversational routines like greetings, invitations, asking for the time, congratulating, wishing luck, etc. in which questions and responses are to some extent predictable from the situations in which they occur (Riemer, 2010, p. 37). However, Chomsky (1959) claims that in the majority of cases the use theory is very complex and finding any regularities or generalisations about the language is hard to achieve. This may be very well illustrated by polysemic words such as *way*, which require a great number of different specific situations to be analysed even for such straightforward uses of *way* as *I don't know the way* or *which way is quicker*?

4.1.3 Mentalist theory

The mentalist theory of meaning constitutes an attempt to explain meaning in terms of what is in people's minds (Kreidler, 1998, p. 43). It is also called the cognitive or conceptual theory of meaning (Filip, 2008).

Foundational works in the cognitive tradition are those by Lakoff and Johnson (1980) and Langacker (1987). Related versions of cognitive semantics can be found in the writings of Jackendoff (1983, 1990), Talmy (1988) and Fillmore (1975, 1976).

Lakoff and Johnson are famous for their ideas of conceptual metaphor and embodied mind. They express the belief that metaphors are conceptual constructions and are central to the development of thought. In their view "our ordinary conceptual system, in terms of which we both think and act, is fundamentally metaphorical in nature" (1980, p. 3). They also argue that the human mind is embodied, by which they mean that most of human cognition depends on and acts using the sensorimotor system and emotions (1999).

Langacker is best known as one of the founders of cognitive linguistics and the originator of Cognitive Grammar, in which language lexicalises concepts by means of phonology, and grammar is the meaningful component linking semantics and phonology (Crystal, 2008, p. 84).

Jackendoff developed a theory of Conceptual Semantics whose goal is to describe how humans express their understanding of the world by means of linguistic utterances. He also created the theory of Parallel Architecture, according to which phonology, syntax, and semantics are autonomous components, each with its own primitives and principles of combination.

Talmy is known for his work on force dynamics, motion, and conceptual structuring. He introduced a theory of force dynamics, a mode describing the way in which entities interact with reference to force, to cognitive linguistics (1981) by claiming that force dynamics goes beyond the traditional notion of the causative. In his view force dynamics is one of the closed-class (grammatical) categories together with recognised categories such as number, aspect or mood. Talmy (1985) also proposed crosslinguistic typologies of lexicalisation patterns especially in the context of their relation to motion events. In his book Toward a Cognitive Semantics (2000), Talmy discusses the issue of the linguistic representation of conceptual structure. He focuses in particular on the interactions between the basic concepts of space and time, motion and location, causation and force interaction, and attention and viewpoint.

Fillmore was one of the founders of cognitive semantics. He developed the theory of Case Grammar (1968), which is concerned with the analysis of the surface syntactic structure of sentences by studying the combination of semantic roles, such as Agent, Object, Benefactor, Location or Instrument required by a specific verb (Crystal, 2008, p. 67). He also elaborated Frame Semantics (1976). In this theory, meanings have an internal structure, which is determined by reference to a background frame or a scene. Currently, Fillmore is involved in the FrameNet project, whose purpose is to create an on-line lexical resource for English, based on frame semantics and supported by corpus evidence.

The core idea shared by all these cognitive approaches to semantics is that all meanings are mental. The theory identifies meaning with concepts, which in Riemer's view (2010, p. 28) are perceived as a way of referring to basic constituents of thought. Language is considered as the 'conduit' for ideas, so words actually mean ideas or concepts. There are, however, a few limitations to this theory. First, while it is quite convincing to say that the meanings of freedom, spelling or love are concepts, it is less obvious for such interjections like 'ouch!', deictic expressions, such as me, you, this or function words, like if, not, like or very. Second, concepts, just like meanings, cannot be seen or identified unambiguously. Riemer (2010, p. 31) points out that certain experiments have been developed to test the properties of particular hypothetical concepts, but this has been done for only a fraction of words and the conclusions are open to a variety of interpretations.

The relation between words, meaning, concepts and objects may be explained using the semiotic triangle presented in Figure 4.3. In the mentalist approach, the three corners of this triangle are word, concept and object.

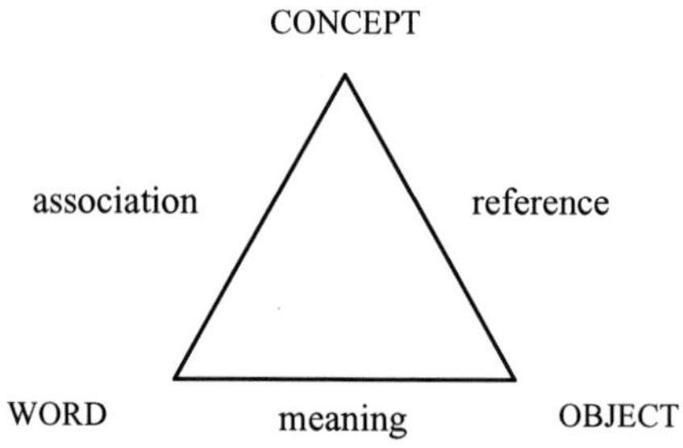

Figure 4-3 Semiotic triangle in the mentalistic theory

The link between word and concept is called an 'association', the link between concept and object 'reference', and the link between object and word 'meaning' (Kreidler, 1998, p. 43). When people hear or read a word, they often form a mental picture of what the word represents, and so they tend to equate a concept with a mental picture. In many cases, however, the meaning of a word is more than what is included in a single image, and human knowledge of these words allows us to do more with them than simply relate them to single objects.

When the semiotic triangle from the mentalist approach (Figure 4-3) is compared with the semiotic triangle from the referential approach (Figure 4-1), the following correspondence between the two variants may be noticed:

concept - thought
word - symbol
object - referent

Concepts are a way of referring to basic constituents of thought. They are expressed and communicated by means of words which are symbols consisting of speech sounds or characters. Concepts refer to objects experienced and encountered in real life such as things, events and situations. Thus, concepts are referents to what the expression is about.

Much of mentalist linguistics is concerned with categorisation. Categorisation is a problematic issue as many words describe concepts which do not have clear category boundaries. Furthermore, not all members of a given category have an equal status (Filip, 2008). Following

Rosch (1975) I can take chair as an example and try to define its meaning in terms of necessary and sufficient conditions. The definition will be as in (24).

(24) a. piece of furniture
b. for one person
c. to sit on
d. having a back and
e. four legs.

I will follow Jackendoff's convention (1983, p. 31) of using capital letters for writing the names of concepts. There are five conditions that the object needs to meet to be recognised as an instance of CHAIR. In fact, this definition is correct for kitchen chairs, but not necessarily for other types of chairs such as dentists' chairs, office chairs, barbers' chairs and beanbag chairs. Kitchen chairs may be considered as the best examples of the category CHAIR. Dentists' chairs are worse examples and beanbag chairs are certainly marginal examples of this category. Rosch (1975) suggests it is more practical to categorise words like CHAIR around good, clear exemplars like kitchen chairs, than to define them in terms of necessary and sufficient conditions. These clear exemplars are prototypes of the category. Their aim is to serve as reference points for the categorisation of not very clear instances. The characteristic feature of the prototype category is that membership of such a category can be graded, e.g. robins and magpies are considered as better examples of BIRD than hummingbirds, ostriches, or penguins.

CHAIR is not the only example of category where some members of a category seem to be better examples of that category than others. A colour category like RED may be analysed instead, as RED has many shades including the red of a fire engine, deep red found on fruit like plums, which may be described as purple, and very pale red described as pink (Riemer, 2010, p. 226). It is not possible to determine a single point on the scale of redness which would be the boundary between red and other colours. Hence, RED cannot be defined by providing necessary and sufficient conditions or anything else that would be a clear category boundary for it. On the other hand, it seems quite apparent that the red of a fire engine is a better example of red than the colour of ripe plums.

A famous example of a category with different statuses of category membership is a series of representations of various cup- and mug-like objects taken from a study by Labov (1973).

Some of the objects in Labov's study are very good examples of cups, and others are very good examples of mugs. There are also a few intermediate cases, in which both descriptions can be applied and cases where none of the labels is suitable.

Wierzbicka (1984, p. 314) rejects a well-established view on categorisation according to which human categorisation of the universe is understood as largely taxonomic (based on hierarchy of kinds). She states that the conceptual relation 'kind of' must be clearly distinguished from the referential relation of set inclusion. Therefore, from a logical point of view it is correct to say all apples are fruit, and not vice versa, but semantically apples are not a kind of fruit. The reason for this is that concepts such as fruit, vegetables, clothing or toys are not taxonomic supercategories in the way that bird is a taxonomic supercategory for swallow or parrot.

Wierzbicka makes a distinction between taxonomic categories, which are prototype-based and non-taxonomic categories, which are not. The former can be illustrated pictorially as they include concepts that are recognised as the best examples of a given category, e.g. oak is a prototypical example of the tree category. The latter cannot be illustrated with pictures, as it is not possible to select the most prototypical example of the given category. For instance, it is hard to judge whether a doll is a better example of the toys category than a trike. Non-taxonomic categories include purely functional concepts, e.g. toys, collectiva-singularia tantum (collective concepts based on contiguity and function), e.g. furniture, collective-pluralia tantum (collective concepts based on contiguity without a reference to function), e.g. left-overs and pseudo-countables (concepts for heterogeneous classes and choppable things), e.g. vegetables (Wierzbicka, 1984, p. 324). Concepts of non-taxonomic categories are fuzzy as they do not identify a particular type of thing but cover all concepts of different kinds which are united by their function and origin.

The applications of conceptual theory will now be considered. The conceptual theory of meaning is also used to explain compositionality and relations between meanings. The concept HORSE-DRAWN CARRIAGE can be analysed into the concepts HORSE and CARRIAGE, and a third element corresponding to the word drawn. The meaning of the linguistic expression horse-drawn carriage also has these three elements. Moreover, they can be changed individually to create different expressions with different meanings, e.g. ox-drawn carriage or horse-drawn plough.

Compositionality cannot be explained fully without referring to the theory of semantic primitives developed by Wierzbicka (1972). Semantic primitives are the basic building blocks of meaning which can be used to construct all other meanings. Most word meanings are not themselves primitive, but are composites of a finite stock of primitive concepts (Riemer, 2010, p. 71). Wierzbicka and Goddard developed a list of semantic primitives, which includes 58 elements which are the simplest possible explanatory terms. They cannot be explained by anything simpler and can be used to create definitions for a large range of concepts. This list includes the following concepts (Goddard, 2002, p. 14): I, you, someone, people, something/thing, body; this, the same, other; one, two, some, all, much/many; good, bad; big, small; think, know, want, feel, see, hear; say, words, true, false; do, happen, move; there is, have; live, die; when/time, now, before, after a long time, a short time, for some time; where/place, here, above, below, far, near, side, inside; not, maybe, can, because, if; very, more; kind of, part of; like.

The conceptual theory also explains meaning relations between words, e.g. hyponymy, which is the relation between a more general and a more specific concept, e.g. CARRIAGE is a member of a wider concept MEANS OF TRANSPORT, and is linked by association with such concepts as COACHMAN, PASSENGER, WHEEL etc. (Jackson, 1988, p. 64). The other type of meaning relation explained by the mentalist theory is synonymy, which depends on the sameness of meaning, e.g. Islamic and Muslim are synonymous because the corresponding concept, which can either be referred to as Islamic or Muslim, is identical for both.

As the theory of meaning relies on the hypothesis that meanings are concepts, it is applicable in communication. People communicate using words, whose meanings are concepts, so they talk about different things, using various concepts, which guarantees genuine communication (Riemer, 2010, p. 30).

Jackendoff (1983, p. 29) opposes the view that the information conveyed by language is about the real world. He makes a clear distinction between the real world (the source of environmental input) and the projected world (experienced world) and claims that people can talk only about their own perception of the world as they have conscious access only to the projected world in their minds. The projected world does not consist of mental images. Experiencing a horse is one thing and experiencing an image of a horse is another. The projected world is unconsciously organised by the mind. Information conveyed by language (reference) must be about the projected world. People can talk about things as long as they have achieved mental representation through these processes of organisation. Abstract entities do not really pose a problem in mentalist semantics as they are not considered more abstract than material entities. The mentalist theory focuses on the projected world, which treats experiencing a house and experiencing love in the same way.

The mentalist theory of meaning seems to be the most widespread of the three discussed above. It is very widely accepted in linguistics and many semanticists consider it as complementary to the referential theory. Its task is bothto identify referents and denotations for the words under examination and to identify concepts with senses and explore conceptual links between the given concept and other concepts (Riemer, 2010, p. 32). I intend to apply this theory to my research and for this reason I will discuss it more fully in this chapter. I will focus particularly on semantic relations and componential analysis.

4.1.4 Semantic relations

Riemer (2010, p. 136) argues that knowing an expression's meaning involves not only knowing its definition but also knowing how it relates to other words in the language. In other words, it requires the establishment of lexical relations such as antonymy, synonymy, meronymy and hyponymy. Such relationships concern the paradigmatic relations of an expression, which determine the choice of a given lexical item over another.

Antonymy is a relationship of incompatibility between two terms with respect to some given dimension of contrast (Riemer, 2010, p. 137). Antonyms of the word *boy* are *girl* and *man*, depending on whether the dimension of contrast is sex or age. There are quite a few expressions which do not have antonyms, e.g. *of* and *corresponding* are cases in which there is no obvious dimension of contrast and, for this reason, antonyms cannot be found. Antonyms can be created morphologically. English forms antonyms productively with the prefix *un-*, e.g. *unexpected* while Polish uses the prefix *nie-* 'not', e.g. *niegrzeczny* 'not polite', i.e. impolite.

Antonyms may be divided into gradable and non-gradable antonyms. Gradable antonyms represent points on a scale, which has a midpoint, e.g. *hot* and *cold* are gradable antonyms as they are two endpoints on a scale, which has a midpoint lexicalised by the adjective *tepid*. Non-gradable antonyms, on the other hand, do not admit a midpoint, e.g. *pass-fail*. Assertion of one of these typically evokes the denial of the other, e.g. *someone who failed the exam did not pass it*.

Synonymy is the relation of meaning identity or meaning similarity (Riemer, 2010, p. 150). Two expressions may be regarded as synonymous if their separately established meanings are identical and if they have the same contextual effect. Synonymy may occur in the forms of sense synonymy and word synonymy. Synonymy of senses is the synonymy of some senses of a word. For example, *pupil* is synonymous with *student* with respect to the sense 'person being instructed by a teacher', but not with respect to the sense 'centre of the eye'. Thus, the two words are synonyms but only with respect to one of their senses.

Word-synonymy is a limiting case of sense-synonymy. It occurs when two words share all their senses. Examples of word synonymy include the pairs Muslim and *Islamic*, *Bombay* and *Mumbai*. In terminology it is assumed that synonymy should be avoided. However, Ullmann (1962, p. 141-142) indicates that word synonymy is common in technical texts as many authors aim for a good style of writing and avoid repetitions by using terms which are mutually substitutable in every context within a particular domain, e.g. in cartography a thematic map portraying area properties using shaded symbols can be synonymously referred to as *choropleth map* and *enumeration map*.

The difference between synonyms very often lies in connotation, which delineates the associations and emotional values of a word. For instance, *violin* and *fiddle* both refer to a musical instrument, but *violin* is the usual term, the neutral one, while *fiddle* is used for humour, or to express affection or lack of esteem (Kreidler, 1998, p. 45).

Meronymy is the relation of the part to the whole, e.g. *hand* is a meronym of *arm*, *seed* is a meronym of *fruit*. The converse relation of whole to part is called holonymy, e.g. *arm* is the holonym of *hand*, *fruit* is the holonym of the seed, etc. The part-whole relationship can be identified by using sentence frames like *X is a part of Y*, or *Y has X* as in *A seed is a part of a fruit*, or *A fruit has seeds* (Saeed, 2003, p. 70). Meronymy reflects the hierarchical classification of lexical items similar to taxonomies. A typical system is presented in Figure 4-4 using the example of *house* which has such meronyms as *roof*, *door*, *floor*, *window*, etc. for which it is a holonym. *Door* consists of *lock*, *handle* and other parts, which are its meronyms.

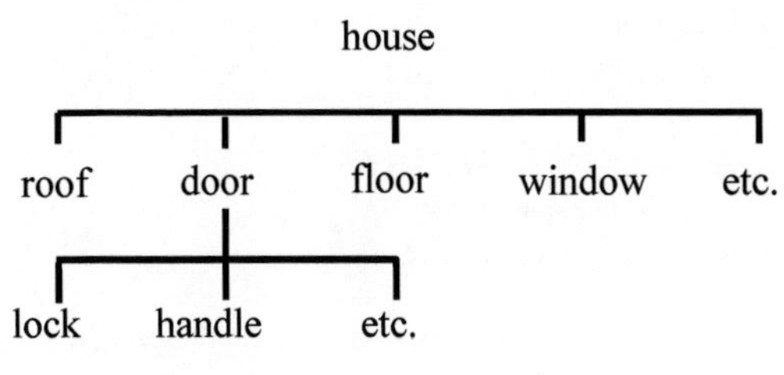

Figure 4-4 Meronymy system after Saeed (2003, p.70)

Meronymy is considered to be transitive, which means that if A is a meronym of B, and B is a meronym of C, then A is also a meronym of C. It may be well illustrated with example (25).

(25) a. *A seed is part of a fruit*;
b. *A fruit is part of a plant*;
c. *A seed is part of a plant.*

The characteristic feature of sentences (25a-c) is the use of *part of*, which is often consistent with the transitivity of the meronymic relation. However, the use of *part of* may be misleading as there are chains of meronymies which are intransitive. When I analyse sentences (26), I realise that *part of* is an indicator of meronymy but is not an indicator of transitivity as the chain of meronymies is both unnatural and false.

(26) a. *Adam's head is part of Adam;*
b. *Adam is part of the Language Department;*
c. *Adam's head is part of the Language Department.*

The examples in (26) are taken from Riemer (2010), who follows Winston et al. (1987). Winston et al. (1987, p. 431) propose a classification of meronymic relations on the basis of the nature of related wholes and parts into six types:

- component-integral object meronymy, e.g. *handle-cup;*
- member-collection meronymy, e.g. *tree-forest;*
- portion-mass, e.g. *slice-pie;*
- stuff-object, e.g. *steel-bike;*
- feature-activity, e.g. *peeling-cooking;*
- place-area, e.g. *Swansea-Wales.*

They argue that meronymy is transitive only when the same type of whole-part relation is involved at the every stage of the chain, as in (26) which contains the place-area type of meronymy as in (27).

(27) a. *Swansea is part of Wales*;
b. *Wales is part of the UK*;
c. *Swansea is part of the UK.*

Another semantic relation is hyponymy, which is described as phrase *kind of/type of/sort of* relation (Riemer, 2010, p. 142). A hyponym is a word containing the meaning of a more general word, e.g. *cat* and *dog* are hyponyms of *animal*; *table*, *stool* and *cupboard* are hyponyms of *furniture*. *Animal* and *furniture*, on the other hand, are hyperonyms or superordinate terms as they include the meaning of more specific words (Riemer, 2010, p. 32).

Hyponymy may be identified on the basis of the notion of class inclusion. Thus, A is a hyponym of B if every A is necessarily a B, but not every B is necessarily an A. For instance, every *car* is a *vehicle*, but not every *vehicle* is a *car* since there are also *trucks*, *vans*, *buses*. In this example, *car* is a hyponym of *vehicle*. Hyponymy is always transitive: if A is a hyponym of B, and B of C, then A is a hyponym of C, e.g. *sports car* is a type of *car*, *car* is a type of *vehicle* and *sports car* is a type of *vehicle*. The semantic relation of hyponymy is presented in Figure 4-5.

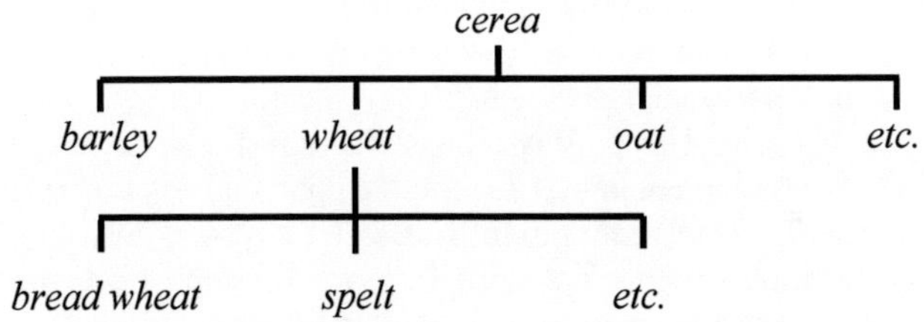

Figure 4-5 The relation of hyponymy after Saeed (2003, p.70)

Wheat is a hyponym of a *cereal* and *bread wheat* is a hyponym of *wheat*. *Cereal* is a hyperonym of *wheat* and *wheat* is hyperonym of *bread wheat*. *Spelt* and *bread wheat* are in a horizontal relation. They are at the same level in the classification hierarchy and they are called co-hyponyms or taxonomic sisters.

A particular type of a system based on hyponymy is taxonomy. According to the *Encyclopaedia Britannica* (2009) taxonomy is the science of biological classification, which is usually restricted to the classification of plants and animals. In taxonomy each hyponym is treated as a strict biological class of the hyperonym (Riemer, 2010, p. 433). What makes taxonomy special, is that different species cover exactly one class. There is a basic template for the taxonomy of plants and animals which was devised by Brown (Saeed, 2003, p. 68). This template consists of five levels: the *unique beginner* or *kingdom* rank (level 0), of which plants and animals are examples. Level 1 is the level of life-forms, e.g. categories like *tree*, *bird*, *fish* in English. Level 2 is generic and includes such items as *oak*, *elm*, *maple*. The generic level may or may not be followed by other levels as, in some taxonomies, it is the last level. Level 3 includes secondary lexemes, which include the term for the superordinate class and a modifier, e.g. *white oak*. Level 4, varietal classes, is very rare as most taxonomies end at the third level. An example of an item that can occur on this level is a *swamp white oak*.

Taxonomy is also a more technical term in classification in general. Its central property is partitioning. It is used, for example, in linguistics to refer to a taxonomic approach, which is an approach to linguistic analysis and description that is predominantly or exclusively concerned with classification (Crystal, 2008).

4.1.5 Componential analysis

Semantic relations of the lexeme contribute to its meaning, which can be studied using a method called componential analysis. The method is a type of definitional analysis which breaks meanings down into binary features (i.e. features with only two possible values, + for truth and - for false). The information contained in componential analysis is similar to the information included in a definition. Basically, anything that can form a part of a definition can be rephrased in terms of semantic components, which are then translated into binary features (Riemer, 2010: 155). Thus, the lexeme may be described through the use of binary semantic components. For example, *sofa*, which is defined as a 'long soft seat with a back and usually arms, on which more than one person can sit at the same time' (Cambridge Advanced Learners' Dictionary, 2011) can be described in componential analysis as [+with back], [+with legs], [-for a single person], [+for sitting], [+with arms], [+rigid]. Componential analysis may be used to provide an entire description of the semantic field on the basis of distinctive binary features. It may be the basis for formulating definitions and establishing semantic relations between the items within the field. Riemer (2010: 155) illustrates componential analysis by using a few concepts from the field of furniture items in English (Table 4-1).

Table 4-1 Componential analysis of English furniture items after Riemer (Riemer, 2010, p. 155)

	with back	with legs	for a single person	for sitting	with arms	rigid
chair	+	+	+	+	-	+
armchair	+	+	+	+	+	+
stool	-	+	+	+	-	+
sofa	+	+	-	+	+	+
beanbag	-	-	+	+	-	-

Among the strong points of componential analysis is the fact that it encourages the assumption that the same distinctive features may be used in different componential analyses, for example, a feature [±edible] used to distinguish *beef* and *cow* may be used to distinguish *plant* and *vegetable*. The method has been criticised widely, however, for its limitations. It is a rigid system, as the only possible values of the semantic feature are + and -. It is suited for a fairly small number of words, mainly nouns with obvious properties which can be transformed into semantic features. It cannot be used for words for which it is difficult to produce traditional definitions, e.g. for colour adjectives. Finally, there are some relational ideas such as the verbs *buy*, *swap*, *sell*, *steal* which can be expressed in the format of traditional definitions but cannot be described using distinctive features, as these ideas have many different interpretations and there are no universal descriptive features that characterise them (Riemer, 2010, p. 157).

When the componential analysis carried out by Riemer (2010) for English furniture items is investigated, it may be noticed that it does not include all senses of some words. For example, *chair* is not only an item of furniture but can also mean 'professorship' and 'head of committee'. In fact, these senses had to be excluded from the componential analysis to make it valid. The word *chair* is an example of a polysemous word, i.e. a word that possesses several distinct senses. Saeed (2003, p. 64) claims that polysemy occurs when the same phonological word has many senses which are judged to be related. The word *chart*, which has two polysemous senses, as illustrated in (28) may be considered.

(28) a. a drawing which shows information in a simple way, often using lines and curves to show amounts;
b. a detailed map of an area of water.

The different senses of *chart* are related. The word was borrowed from the French word *carte* 'card, map', but it originally comes from the Latin *carta* 'paper, card, map'. It was borrowed in the 16th century when it became an accepted term for 'map'. The contemporary meanings of the word *chart* developed through the processes of semantic extension. Lexicographers deal with polysemy by listing polysemous senses under one entry.

Polysemy is often contrasted with homonymy, which occurs where a single phonological form has unrelated meanings, e.g. the English verb [weiv] is spelt *wave* or *waive*, depending on the meaning (Riemer, 2010, p. 161). The verb *wave* derives from Old English *vafian* and means 'to raise your hand and move it from side to side as a way of greeting someone, telling them to do something or adding emphasis to an expression', while the word *waive* was borrowed into English from Old French *gaiver* and means 'to not demand something you have a right to, or not cause a rule to be obeyed'. *Waive* and *wave* are homophones, which are types of homonyms based on unrelated senses of the same spoken word. There is a second group of homonyms, homographs, which have unrelated senses and different pronunciation, e.g. lead ('metal') and (to) lead (Saeed, 2003, p. 63).

4.2 Ontologies

The word *ontology* has different senses in different communities. The primary difference is between the philosophical sense and the computational sense (Guarino, Oberle & Staab, 2009, p. 1). This basic distinction has consequences for the spelling of the word ontology. Written with an uppercase initial letter *Ontology* refers to a philosophical discipline, which deals with the nature and structure of things. Combined with an indefinite article and written with a lowercase initial it is used mainly in Computer Science to indicate a special kind of information or computational artefact. The term was then adopted in Artificial Intelligence (AI) in the 1980s to refer to both a theory of a modelled world and a component of knowledge systems (Gruber, 2009).

The notion of *ontology* was defined by Gruber (1993) as a specification of a conceptualisation used to help humans and computers share knowledge. The conceptualisation is understood as an abstract, simplified view of the world that people wish to represent for some purpose. The specification is the representation of this conceptualisation in a concrete form. A slightly different definition was proposed by Borst (1997). In his view an ontology is a formal specification of a shared conceptualisation. The notion of the shared conceptualisation implies that the conceptualisation should express a view shared by several parties. Studer et al. (1998) merged the two definitions into one which states that an ontology is a formal, explicit specification of a shared conceptualisation. This means that an ontology specifies the concepts, relationships, and other distinctions that are relevant for modelling a domain. Jackendoff (1983) does not agree with the idea of the shared conceptualisation. In his view, every single individual has their own conceptualisation. The real world is a shared object but the projected world (experienced world) is different for everyone as people have different perception of the real world and different conceptualisations.

Ontologies, just like databases, are knowledge organisation systems. Stevens et al. (2000) argue that the main difference between the two is that a database schema is generally not reusable, while an ontology can be re-used as a true model of a portion of the world. This claim may be true for some databases, particularly old ones which are available in a printed version only. More recent databases created in some kind of software, e.g. MS Access, may be used in a number of different applications, as their structure can be easily modified, e.g. by changing the type of data in the fields, by creating nested tables, by adding and removing columns or by merging various databases. Electronic databases can now be exported and saved in different formats so that they can be further reused, for example as Internet applications.

The actual difference between a database and an ontology is the way they approach semantics. While databases, in particular termbases, typically keep the semantics implicit and offer a traditional representation of concepts and their relationships, usually in some sort of table, ontologies allow one to build domain models and show the relationships between concepts more explicitly by taking advantage of diagrams. Databases have two distinct parts: data and formalism. Ontologies, on the other hand, constitute one integrated whole.

Ontologies are created using data collected in termbases which contain information on concept systems but represent a more refined method for presenting and sharing these data.

Ontologies nowadays are connected inseparably with software and database engineering. Therefore, before I take this discussion any further it is necessary to make a basic distinction between ontologies in the sense of software that is used to instantiate knowledge organisation systems (4.2.1) and ontologies in the sense of knowledge organisation systems (4.2.2). For the former I will discuss software packages for creating ontologies such as

CAOS or Protégé, and for the latter I will deal with off-the-shelf ontologies such as GOLD or WordNet.

4.2.1 Software for creating ontologies

Software for creating ontologies facilitates building a domain model and specifying its representation. The user working with the software takes a text in a particular domain as a starting point and when reading through this text he/she identifies concepts and establishes relations between them. The user may also start from a collection of terms which has been prepared already, for example by a terminologist, and look for concepts and conceptual links between them. He/she uses the software as a tool to build the concept system of this particular domain.

State-of-the-art software for ontologies facilitates a graphic representation of concept systems in the form of diagrams. It also supports reusability of data so that the data can be exported, translated, queried and unified across independently developed systems and services (Gruber, 2009). Software packages use representation languages independent of data modelling strategy or implementation to formulate concept systems. The current W3C Semantic Web standard for encoding ontologies is OWL (Web Ontology Language).

As an example of the software package used to build ontologies, I will discuss CAOS. CAOS stands for a Computer-Aided Ontology Structuring. It is a tool for constructing terminological ontologies, which are domain-specific ontologies that model concepts and the relations between them. They clarify and define concepts within the particular domain and may be used for translation purposes to establish equivalence relations between concepts in different languages. Terminological ontologies differ from other ontologies in the fact that the latter deal with the world in general and rely on general language, while terminological ontologies focus on a narrow, specialised field of this world and operate on specialised terminology used within this field (Madsen & Thomsen, 2009).

CAOS is a joint research project by Bodil Nistrup Madsen, Hanne Erdman Thomsen and Carl Vikner from the Department of Computational Linguistics at Copenhagen Business School. The first version of CAOS was elaborated between 1998 and 2002 and the work on the current, second version of CAOS started in 2005 (Madsen, Thomsen & Vikner, 2006). CAOS 2 has been enhanced by adding a graphical presentation of ontologies and some work has been done to implement different types of concept relations. CAOS is designed to be interactive and quite straightforward to use as it presupposes that the end-user has a background in terminology rather than in formal ontology (Madsen, 2007, p. 182).

CAOS 2 is a single-user programme. There is no network database integrated with this application, so ontologies created by a user have to be stored on the user's local hard disk or network profile. CAOS 2 runs in a Java Environment. Access to CAOS 2 is limited as it requires a username and password which may be obtained only by contacting the CAOS team directly.

CAOS 2 consists of two parts: the database and the diagram. The database is saved locally when saving projects. The diagram, which reflects the ontology architecture, is copied from the database structure. The user can modify the ontology by operating on the diagram boxes.

An ontology constructed using CAOS consists of concepts, which are building blocks, and relations, which glue concepts together. Relations are divided into three categories:

- generic relations or type relations, which are directed from a conceptual supertype (e.g. *flower*) to a subtype (e.g. *rose*);
- part-whole relations, which are directed from the whole (e.g. *car*) to the part (e.g. *wheel*);
- associative relations, which relate concepts across tree structures.

Associative relations are specified for verbs or verb phrases, often in conjunction with a preposition. Associative relations represent, for example, the functions and processes the concept is involved in, e.g. Protein has Function Receptor, Protein is associated with process transcription and Protein has organism classification species. In this record, concepts are given separately and types of relations are written together in the machine font (Stevens, Goble & Bechhofer, 2000). Associative relations may be found in a definition of the concept *dividend* given in (29).

(29) dividend – a taxable payment declared by a company's board of directors and given to its shareholders out of the company's current or retained earnings quarterly. (InvestorWords, 2011).

Associative relations 'tell the whole story' of the word *dividend* excluding the fact that it is a type of payment (which is reflected by a generic relation).

It is possible to use all types of concept relations in CAOS 2. The user can define and introduce their own relations to the system as well (Madsen & Thomsen, 2009, p. 279). CAOS 2 uses concept diagrams. Diagramming methods are based on UML (Unified Modelling Language). The classes (boxes) of UML are used to model concepts. The upper part of a box contains the term representing the concept and a systematic number which describes the position of the concept in a concept system. The middle part contains dimension specifications, written in bold and in uppercase, and the lower part features specifications.

Characteristics that may be attributed to concepts in CAOS are called feature specifications. They consist of attributes and their values and are the backbone of terminological concept modelling. A concept automatically inherits all the feature specifications of its superordinate concepts. An attribute in a feature specification whose possible values allow a distinction between some of the subconcepts of the given concept is called a dimension of a concept. A dimension specification contains a dimension and the values associated with the corresponding attribute in the feature specifications of the subordinate concepts: 'DIMENSION: [value1 | value2, …]'. One or more dimensions of a concept may play the role of subdividing dimensions, i.e. dimensions that are used in definitions of the concept's closest subconcepts.

The idea of CAOS is that the terminologist working on a particular domain reads texts in this domain and builds the ontology by entering concept relations, feature specifications and dimension specifications. Madsen (2007, p. 184) describes how the ontology for printers presented in Figure 4-6 was built. The terminologist used a text on the classification of printers found in an introduction to elementary computer science. The text specifies such features of printers (dimensions in CAOS terminology) as CHARACTER TRANSFER, COPY and NOISE. The terminologist has to specify attributes for the dimensions, which are as follows 'CHARACTER TRANSFER: [impact | nonimpact]', 'COPY: [multiple | single]' and 'NOISE: [noisy | quiet]'. Then, he/she selects CHARACTER TRANSFER as a subdividing dimension because other characteristics are connected with it. The subdividing dimensions are shown in bold characters. This action will cause CAOS to divide printers into subconcepts: *impact printer* and *nonimpact printer* and will ask the terminologist how to assign other dimension values (i.e. which attributes with values to which subconcept). Then, it will generate the diagram with 1.1 and 1.2 as shown in Figure 4-6.

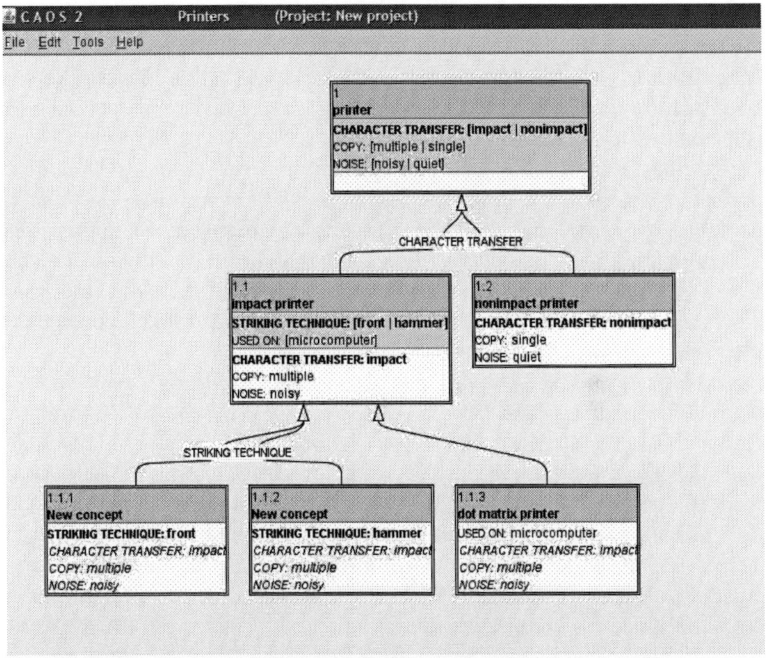

Figure 4-6 Draft concept system in CAOS. Based on Madsen (2007, p. 184). Reprinted with kind permission from John Benjamins Publishing Company, Amsterdam/Philadelphia. [www.benjamins.com].

The terminologist is able to specify dimensions of impact printer on the basis of the text that he/she is reading as STRIKING TECHNIQUE and USED ON. The former dimension is selected as a subdividing dimension. This will cause CAOS to create two subconcepts with the feature specifications: 'STRIKING TECHNIQUE: front' and 'STRIKING TECHNIQUE: hammer'. The concepts in position 1.1.1 and 1.1.2 have not been given any designations as the terminologist could not find them in the documentation. The idea is that they will be found by going through other texts in this domain later on. CAOS creates a third subconcept in position 1.1.3 by assigning the feature specification 'USED ON: microcomputer' directly to the third subconcept of impact printer and asks the terminologist to enter the designation. The terminologist ascribes the designation dot matrix printer to the concept. The feature specification assigned directly to a given concept is called a primary feature specification in contrast to an inherited feature specification, which is inherited from the concept's superordinate concepts. Thus, all feature specifications in dot matrix printer apart from 'USED ON: microcomputer' are inherited from the concept impact printer. The primary feature specifications in CAOS are given in ordinary type, while inherited feature specifications are in italics. The box in a red frame is the one that has been selected by the terminologist to work with.

The basic assumption of CAOS is that a single ontology should not contain indeterminacy. Madsen (2007, p. 187) claims that this can be achieved by complying with the principle of uniqueness of dimensions. The principle states that a given dimension may occur on only one concept in an ontology. Thus, primary feature specifications with the same attribute must always occur in sister concepts, i.e. coordinate concepts on the same level which have the same superordinate concept.

Protégé is another software package for the creation of ontologies. It was developed by the Stanford Center for Biomedical Informatics Research at the Stanford University School of Medicine. Its advantage over CAOS, which is free only to academics by obtaining a user account from the CAOS team, is that it is open to the public and may be downloaded and used at any time and does not require a user account. I-Term, which is an alternative to CAOS created at the Copenhagen Business School, is available free of charge in the demo version only. It was developed for commercial purposes and the cost of the full version is €10,000 for a basic licence for 10 users (version 3.4.9 of the software).

The Protégé platform supports two main ways of modelling ontologies: via the Protégé-Frames and Protégé-OWL editors (Protégé, 2010). Protégé-Frames editor supports building and populating ontologies that are frame-based, complying with the rules of the Open Knowledge Base Connectivity protocol (OKBC). A frame-based ontology in Protégé consists of a "set of classes organised in a subsumption hierarchy to represent a domain's salient concepts, a set of slots assigned to classes to describe their properties and relationships, and a set of instances of those classes - individual exemplars of the concepts that hold specific values for their properties" (Protégé, 2010).

The Protégé-OWL editor enables users to build ontologies for the Semantic Web, in particular in the W3C's Web Ontology Language (OWL). An OWL ontology may include descriptions of classes, properties and their instances.

Protégé ontologies can be exported into a variety of formats including RDF(S), OWL, and XML Schema. Protégé, just like CAOS, is based on Java. It may be extended and provides a plug-and-play environment which makes it a flexible basis for quick prototyping and application development.

4.2.2 Ontologies as knowledge organisation systems

This section concerns ontologies as knowledge organisation systems. I will look mainly at ontologies for general language. As an example of general ontology, I will investigate WordNet and its derivatives: EuroWordNet and Polish WordNet. I will also consider how terminological ontologies have been built, using GOLD ontology as an example.

WordNet is one of the most successful ontologies for general language available for English. It has been in use since 1985 (Riemer, 2010, p. 272) and was developed under the direction of George A. Miller (Fellbaum, 1998). It is as an electronic lexical database of English, containing nouns, verbs, adjectives and adverbs, which is freely and publicly available for download and is particularly designed for those working in the field of computational linguistics and natural language processing. Its aim is to represent and organise lexical semantic data in a form which optimises information retrieval. The main organisation unit in WordNet is the *synset* (Riemer, 2010, p. 272). Synsets, conventionally marked by curly brackets, group near-synonyms, like {beat, crush, trounce, vanquish}, which identify a particular lexicalised concept, e.g. the 'defeat' sense of *beat*. Each synset is provided with a short definition or gloss. For example, the WordNet gloss of the given synset is 'come out better in a competition, race, or conflict'. The gloss works as a definition of all the words in a synset. Each synset expresses a distinct concept. Synsets are linked together by means of conceptual-semantic and lexical relations. The resulting network of meaningfully related concepts can be navigated through the browser.

WordNet deals with polysemy by assigning polysemous words to different synsets, one for each sense. Thus the verb *beat* also belongs to such synsets as: {beat, flap}, {beat, scramble}, {beat, tick, ticktock, ticktack} and twenty others (Princeton University, 2010). Each of these synsets reflects a different sense of the verb – the move with a flapping motion' sense, the 'stir vigorously' sense, etc. There are also a few senses of the word *beat* where no synonyms are available. In such situations, a gloss of the meaning is provided to identify the intended sense, e.g. {beat (produce a rhythm by striking repeatedly)}. The word *beat* also occurs as a noun in the following synsets: {pulse, pulsation, heartbeat, beat} meaning 'the rhythmic contraction and expansion of the arteries with each beat of the heart' or {rhythm, beat, musical rhythm} meaning 'the basic rhythmic unit in a piece of music'. The latter synset includes a multi-word unit *musical rhythm*. Finally, it may also be an adjective, as is documented in the synset {all in, beat, bushed, dead}, meaning 'very tired'.

Another lexical relation included in the WordNet structure is named 'hyponymy/taxonomy'. The database has the facility of displaying the whole hyponymic/taxonomic hierarchy for every noun. For example, *table* and *furniture* are linked by the hyponymy relation with *furniture* being the superordinate term. The WordNet entry for *table* includes a pointer to *furniture*, thus allowing the user to see a superordinate term. This superordinate is called *hypernym* in WordNet terminology (more commonly known as hyperonym). *Hypernymy* is one of the four semantic relations which can be displayed in WordNet. Figure 4-7 includes links to three other relations, which are *hyponymy* (direct hyponymy and full hyponymy), *part meronymy* (more commonly known as meronymy) and *has instances.*

Noun

- S: (n) **table**, tabular array (a set of data arranged in rows and columns) *"see table 1"*
- S: (n) **table** (a piece of furniture having a smooth flat top that is usually supported by one or more vertical legs) *"it was a sturdy table"*
  - *direct hyponym* / *full hyponym*
  - *part meronym*
  - *has instance*
  - *direct hypernym* / ***inherited hypernym*** / *sister term*
    - S: (n) furniture, piece of furniture, article of furniture (furnishings that make a room or other area ready for occupancy) *"they had too much furniture for the small apartment"; "there was only one piece of furniture in the room"*
      - S: (n) furnishing ((usually plural) the instrumentalities (furniture and appliances and other movable accessories including curtains and rugs) that make a home (or other area) livable)
        - S: (n) instrumentality, instrumentation (an artifact (or system of artifacts) that is instrumental in accomplishing some end)
          - S: (n) artifact, artefact (a man-made object taken as a whole)
            - S: (n) whole, unit (an assemblage of parts that is regarded as a single entity) *"how big is that part compared to the whole?"; "the team is a unit"*
              - S: (n) object, physical object (a tangible and visible entity; an entity that can cast a shadow) *"it was full of rackets, balls and other objects"*
                - S: (n) physical entity (an entity that has physical existence)
                  - S: (n) entity (that which is perceived or known or inferred to have its own distinct existence (living or nonliving))

Figure 4-7 Representation of the noun *table* in WordNet (Princeton University, 2010)

I will start discussing semantic relations in WordNet from *hypernymy* as it is the most complex relation. After choosing the synset that refers *table* to *furniture*, the user can select the type of semantic relation to be displayed for a given synset. After selecting '*direct hypernym*' the synset {furniture, piece of furniture, article of furniture} is presented. Selecting '*inherited hyperonyms'* displays results in rendering the whole class of inheritance for *furniture*, which includes such sysnsets as {furnishing}, {instrumentality, instrumentation}, {artifact, artefact}, etc. The most recently selected type of relation to be presented is highlighted in bold, just like *inherited hypernym* in Figure 4-7.

WordNet also provides information on sister terms (which include all types of furniture), e.g. *cabinet, chest of drawers,* direct hyponyms, which include all possible types of tables, e.g. *altar*, *coffee table*, *stand, desk*, *worktable* and full hyponyms, which include subtypes of tables, e.g. *worktable*, has such hyponyms as *drafting table* and *workbench* and a *laboratory bench* is a type of *workbench.*

As the next semantic relation I will discuss *hyponymy.* WordNet shows *direct hyponyms,* which include such types of tables as *coffee table*, *desk*, *counter*, *stand* and *full hyponyms*, which provide all subtypes of *tables* specified in the *direct hyponym* class, e.g. *desk* has such subtypes as *davenport* (a small decorative writing desk), *secretary* and *writing desk.*

Two other relations available in WordNet are *part meronyms*, e.g. *leg*, *tabletop* and *has instances*, which is a very small group including two items only: *Round Table* and *King Arthur's Round Table*. It is important to note that this last class includes proper names written with capital letters and is also linked to hyponyms as *Round Table* or *King Arthur's Round Table;* included among tables just like *kitchen table* or *tea table*.

If, for my analysis, I had selected a more general word, such as *furniture,* WordNet would have additionally displayed domain term category, e.g. *Americana* 'any artifact (such as books or furniture or art) that is distinctive of America' or *rosemaling* 'a Scandinavian style of carved or painted decoration (as on furniture or walls or dinnerware) consisting of floral motifs' and derivationally related forms, e.g. *furnish.* Antonyms can also be displayed if entries have them, e.g. the antonym for *woman* is *man*.

The WordNet structure features inheritance hierarchy as every noun is linked to its hyperonyms and hyponyms. The WordNet user can see the complete inheritance hierarchy with all hyperonyms, hyponyms, meronyms and sister concepts at once and, therefore, can access more definitional information than would be available for the single term only (Riemer, 2010, p. 275).

The representation of verbs in WordNet is similar to the representation of nouns. It is also organised by inheritance hierarchy, but verb hierarchies are much shallower than noun hierarchies and typically reach only four levels (Fellbaum, 1990, p. 287). In the case of verbs, WordNet provides not only semantic relations but also lexical ones. The entry for *walk* presented in Figure 4-8 has eight links to different types of data. These are *troponym*, *verb group*, *hypernym*, *entailment*, *phrasal verb*, *antonym*, *derivationally related form* and *sentence frame*. Out of these eight types of information, only hypernymy, antonymy and derivationally-related forms are shared by verbs and nouns. WordNet provides hyperonyms and sister terms for verbs, e.g. the verb *walk* has a hyperonym *travel* and a number of sister terms, e.g. *go around*, *come*. Derivationally-related forms include *walker* and *walking*, whereas the antonym for *walk* is *ride*. Additional information in verbal entries and not present in nominal entries incorporates sentence frames, e.g. *Somebody walks*, phrasal verbs, e.g. *walk around*, an entailment relation, e.g. *step*. The verb group class covers different meanings of the entry, e.g. *walk* as *traverse or cover by walking*, *walk* as *accompany or escort*, *walk* as *take the air* and *walk* as *make walk*. There is also a semantic relation of troponymy, which is peculiar to verbs. The notion of troponymy was proposed as a new concept by Fellbaum and Miller (1990) and it applies to a verb which elaborates the specific manner in which another verb is performed, e.g. *march*, *pace*, *stride*. It plays a similar role to that of hyponymy in nouns.

Adjectives in WordNet are arranged in clusters which contain head synsets and satellite synsets. Head synsets usually have a few satellite synsets. Each of these satellite synsets represents a concept similar to the one represented by the head synset (Princeton University, 2010). Each of the synsets contains antonyms and those synsets that have similar meaning to the given synset. Derivationally-related forms and attributes may also be given. Figure 4-9 provides a view for the adjective *intelligent*. This view was arrived at in a similar way to the view in Figure 4-7.

Verb

- S: (v) **walk** (use one's feet to advance; advance by steps) *"Walk, don't run!"; "We walked instead of driving"; "She walks with a slight limp"; "The patient cannot walk yet"; "Walk over to the cabinet"*
    - *direct troponym / full troponym*
    - *verb group*
        - S: (v) **walk** (traverse or cover by walking) *"Walk the tightrope"; "Paul walked the streets of Damascus"; "She walks 3 miles every day"*
        - S: (v) **walk** (accompany or escort) *"I'll walk you to your car"*
        - S: (v) **walk**, take the air (take a walk; go for a walk; walk for pleasure) *"The lovers held hands while walking"; "We like to walk every Sunday"*
        - S: (v) **walk** (make walk) *"He walks the horse up the mountain"; "Walk the dog twice a day"*
        - S: (v) **walk** (walk at a pace) *"The horses walked across the meadow"*
    - *direct hypernym / inherited hypernym / sister term*
    - *entailment*
        - S: (v) step (shift or move by taking a step) *"step back"*
    - *phrasal verb*
        - W: (v) walk around [Related to: walk] (walk with no particular goal) *"we were walking around in the garden"; "after breakfast, she walked about in the park"*
        - W: (v) walk about [Related to: walk] (walk with no particular goal) *"we were walking around in the garden"; "after breakfast, she walked about in the park"*
    - *antonym*
    - *derivationally related form*
    - ***sentence frame***
        - Something ----s
        - Somebody ----s
        - Somebody ----s PP
        - [Applies to walk] Sam and Sue walk
        - [Applies to walk] The children walk to the playground

Figure 4-8 Representation of the verb *walk* in WordNet (Princeton University, 2010)

Adjective

- S: (adj) **intelligent** (having the capacity for thought and reason especially to a high degree) *"is there intelligent life in the universe?"; "an intelligent question"*
  - *see also*
    - S: (adj) precocious (characterized by or characteristic of exceptionally early development or maturity (especially in mental aptitude)) *"a precocious child"; "a precocious achievement"*
    - S: (adj) smart (showing mental alertness and calculation and resourcefulness)
  - *similar to*
    - S: (adj) agile, nimble (mentally quick) *"an agile mind"; "nimble wits"*
    - S: (adj) apt, clever (mentally quick and resourceful) *"an apt pupil"; "you are a clever man...you reason well and your wit is bold"-Bram Stoker*
    - S: (adj) brainy, brilliant, smart as a whip (having or marked by unusual and impressive intelligence) *"our project needs brainy women"; "a brilliant mind"; "a brilliant solution to the problem"*
    - S: (adj) bright, smart (characterized by quickness and ease in learning) *"some children are brighter in one subject than another"; "smart children talk earlier than the average"*
    - S: (adj) natural, born, innate (being talented through inherited qualities) *"a natural leader"; "a born musician"; "an innate talent"*
    - S: (adj) quick, ready (apprehending and responding with speed and sensitivity) *"a quick mind"; "a ready wit"*
    - S: (adj) prehensile (having a keen intellect) *"poets--those gifted strangely prehensile men"- A.T.Quiller-Couch*
    - S: (adj) scintillating (brilliantly clever) *"scintillating wit"; "a play full of scintillating dialogue"*
    - S: (adj) searching, trenchant (having keenness and forcefulness and penetration in thought, expression, or intellect) *"searching insights"; "trenchant criticism"*
  - ***attribute***
    - S: (n) intelligence (the ability to comprehend; to understand and profit from experience)
  - *antonym*
  - *derivationally related form*

Figure 4-9 Representation of the adjective *intelligent* in WordNet (Princeton University, 2010)

It is important to note that, apart from qualitative adjectives such as the one presented above, WordNet may also include relational adjectives, which are not a part of adjective cluster organisation and do not have antonyms. Relational adjectives include so-called *pertainyms*. A pertainym is a lexical pointer to the form, the searched word is 'pertaining to'. Relational adjectives pertain to nouns, e.g. *cadastral* is a pertainym of the noun *cadastre*. The synset for a relational adjective typically contains only one word or collocation, a pertainym and a derivationaly related form.

Adverbs in WordNet are often derived from adjectives, and sometimes have antonyms. The synset for an adverb usually contains a pertainym, which points to the adjective from which it is derived, e.g. the adverb *quickly* points to the adjective *quick*.

WordNet is not free of some deficiencies. Riemer (2010, p. 726) criticises WordNet entries for not showing the fundamental differences between homonymy and polysemy. The user of WordNet who consults the entry for *vessel*, will find no indication that 'a tube in which a body fluid circulates' and 'a craft designed for water transportation' are not semantically related. The senses of the polysemous word *trouble*, which include 'problem' and 'hassle' are presented in the same way as senses of the word *vessel*.

Another point of criticism concerns the fact that WordNet does not recognise syntagmatic relations between words like *ball*, *racket* and *net*, which belong to the same domain of ball games. The contextual relation between the three cannot be presented in any way in WordNet, as it does not correspond to any of the semantic relations recognised by the database. Riemer (2010, p. 277) blames the WordNet architecture for this shortcoming and claims that it limits psycholinguistic evidence. Riemer's argument does not seem to be justified, however. Wordnet is useful as it attempts to organise the psycholinguistic evidence in a structured way.Furthermore, its architecture could be modified by introducing a new named relation that would reflect the syntagmatic relations between the objects belonging to the same domain.

The original WordNet developed at Princeton University triggered the EuroWordNet (EWN) project, in the framework of which a multilingual database with wordnets for several European languages was created (Vossen, 1998, p. 3). The project took three years from 1996 to 1999 and was carried out in two stages. In the first stage a database for English, Dutch, Spanish and Italian was created for 30,000 concepts (50,000 word meanings). In the second stage, German, French, Czech and Estonian were added and the database was supplemented with an additional 15,000 concepts (25,000 word meanings). To provide some level of comparison, the original WordNet covers over 117,000 concepts (synsets) and over 150,000 English words (Vossen, 2009).

EuroWordNet added features to relations (Vossen, 2009). For example, in EWN the *entailment relation* has been divided into *has subevent* and *in manner* (Piasecki, Szpakowicz & Broda, 2009). It also introduced *Cross-Part-Of-Speech relations*, such as *manner*, e.g. *to slurp – noisily*.

The wordnets in different languages in EuroWordNet were linked to the Inter-Lingual-Index, based on the Princeton WordNet. The index connects the languages so that it is possible to go from synsets in one language to similar synsets in any other language. Index-records are based on WordNet synsets and include synonyms, glosses and source references. Each meaning is linked with an equivalence relation to a WordNet synset (Vossen, 1997).

EuroWordNet is a very useful application and constitutes a first attempt at combining wordnets in different languages. However, it is not free of deficiencies. Not all wordnets can communicate with one another as they are linked to different versions of the English wordnet. Free access and usage are restricted by proprietary rights (Vossen, 2009).

After the EuroWordNet project had finished, Global Wordnet Association was founded in 2000 to maintain the framework. Currently, wordnets exist for more than 50 languages, including Arabic, Bantu, Basque, Chinese, Bulgarian and Polish (Vossen, 2009).

Work on the Polish WordNet started in 2005 and was undertaken by the Institute of Applied Mathematics of Wrocław University of Technology. The project lasted for three years and was completed in 2008. The size of the present state of the Polish WordNet – version 1.0 is 17,695 synsets (concepts) and 26,990 lexical units (meanings) (Polish WordNet, 2010).

The structure of the Polish WordNet (called *plWordNet* by its creators) follows the model of the Princeton WordNet and the structures of the EuroWordNet. The creators of plWordNet have not localised the English WordNet but constructed the lexical network from scratch in two stages. In the first stage, the linguistic principles were established including a list of semantic relations with detailed diagnostic tests. A client software tool that records the lexicographers' decisions in a central database was also implemented. A core WordNet was then populated with around 10,000 of the most frequent lexemes in the IPI PAN (Instytut Podstaw Informatyki Polskiej Akademii Nauk 'Institute of Computer Science at the Polish Academy of Sciences') Corpus (Derwojedowa et al., 2007).

In the second phase, the software detected candidate semantic relations in the corpus, using statistical methods of grouping words by semantic similarity. These relations were then suggested to linguists/lexicographers via a graphical user interface for revision and approval (Wroclaw University of Technology, 2006).

Polish WordNet describes its synsets in terms of such semantic relations as synonymy, antonymy, hyperonymy/hoponymy and meronymy/ holonymy. It also covers cross-categorial relations (relatedness, pertainymy and fuzzynymy) which model word formation relations. Piasecki et al. (2009, p. 32-35) claim that only relatedness is specific to Polish WordNet and pertainymy was adopted from the original WordNet and fuzzynymy from EuroWordNet. However, this last concept is hard to localise in literature and the creators of Polish WordNet do not provide bibliographic information on its origin. It is important to note that the three relations were broadened for Polish as the Polish language differs from English in many aspects including morphology and grammar. Therefore, they obtain quite a different interpretation in Polish.

Relatedness applies not only to relational adjectives and their nominal base words but also to other derivational classes such as deadjectival nouns, e.g. *biały* 'white': *białość* 'whiteness', aspectual verbs, e.g. *czytać* 'to read$_{\mathrm{IMPERF}}$' : *przeczytać* 'to read$_{\mathrm{PERF}}$' and causatives derived from adjectives, e.g. *zasmucić kogoś* 'make$_{\mathrm{PERF}}$ someone sad': *smutny* 'sad' (Derwojedowa *et al.*, 2007).

Pertainymy covers frequently occurring but less regular derivational relations such as people, establishments and their central attributes, e.g. *księgarz* 'bookseller' : *księgarnia* 'bookstore' : *książka* 'book', derived feminine forms, e.g. *aktor* 'actor' : *aktorka* 'actress' and names of young animals and other dimunitive and augmentative names, e.g. *pies* 'dog' : *piesek* 'little dog' : *psisko* 'big dog' (Derwojedowa *et al.*, 2007).

Finally, fuzzynymy connects pairs of LUs which are connected by syntagmatic relations, but which cannot be fitted into the existing system of relations. Fuzzynymy links nouns that describe employees and their workplaces, e.g. *listonosz* 'postman' and *poczta* 'post office', names of objects and places where these objects are located, e.g. *roślina* 'plant' and *ogród* 'garden' or activities and places where they take place, e.g. *spacerowanie* 'walking' and *park* 'park' (Piasecki, Szpakowicz & Broda, 2009, p. 34). The entry for *krzesło* 'chair' in Polish WordNet is shown in Figure 4-10 and the translation of the entry may be found in Appendix 4. Figures in the brackets indicate a number of relations within a given relation type, e.g. (2/3) for hiperonimia 'hyperonymy' means that there are two hyperonyms (*krzesło 1* and and *krzesło 2*) and three hyponyms (*krzesłko 1*, *stołeczek 2* and *krzesło wiedeńskie 1*).

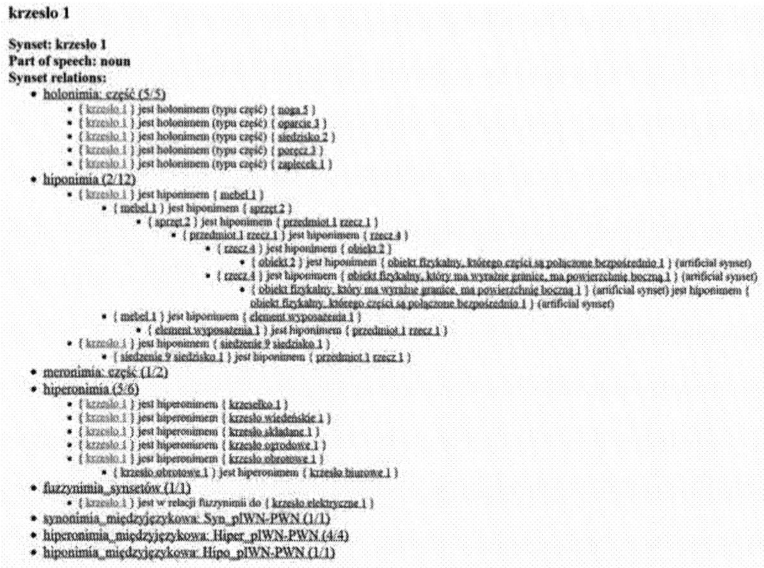

krzeslo 1

Synset: krzeslo 1
Part of speech: noun
Synset relations:

- holonimia: część (5/5)
    - { krzeslo.1 } jest holonimem (typu część) { noga.5 }
    - { krzeslo.1 } jest holonimem (typu część) { oparcie.3 }
    - { krzeslo.1 } jest holonimem (typu część) { siedzisko.2 }
    - { krzeslo.1 } jest holonimem (typu część) { poręcz.1 }
    - { krzeslo.1 } jest holonimem (typu część) { zapiecek.1 }
- hiponimia (2/12)
    - { krzeslo.1 } jest hiponimem { mebel.1 }
        - { mebel.1 } jest hiponimem { sprzęt.2 }
            - { sprzęt.2 } jest hiponimem { przedmiot.1 rzecz.1 }
                - { przedmiot.1 rzecz.1 } jest hiponimem { rzecz.4 }
                    - { rzecz.4 } jest hiponimem { obiekt.2 }
                        - { obiekt.2 } jest hiponimem { obiekt fizykalny, którego części są połączone bezpośrednio.1 } (artificial synset)
                    - { rzecz.4 } jest hiponimem { obiekt fizykalny, który ma wyraźne granice, ma powierzchnię boczną.1 } (artificial synset)
                        - { obiekt fizykalny, który ma wyraźne granice, ma powierzchnię boczną.1 } (artificial synset) jest hiponimem { obiekt fizykalny, którego części są połączone bezpośrednio.1 } (artificial synset)
        - { mebel.1 } jest hiponimem { element wyposażenia.1 }
            - { element wyposażenia.1 } jest hiponimem { przedmiot.1 rzecz.1 }
    - { krzeslo.1 } jest hiponimem { siedzenie.9 siedzisko.1 }
        - { siedzenie.9 siedzisko.1 } jest hiponimem { przedmiot.1 rzecz.1 }
- meronimia: część (1/2)
- hiperonimia (5/6)
    - { krzeslo.1 } jest hiperonimem { krzesełko.1 }
    - { krzeslo.1 } jest hiperonimem { krzesło wiedeńskie.1 }
    - { krzeslo.1 } jest hiperonimem { krzesło składane.1 }
    - { krzeslo.1 } jest hiperonimem { krzesło ogrodowe.1 }
    - { krzeslo.1 } jest hiperonimem { krzesło obrotowe.1 }
        - { krzesło obrotowe.1 } jest hiperonimem { krzesło biurowe.1 }
- fuzzynimia_synsetów (1/1)
    - { krzeslo.1 } jest w relacji fuzzynimii do { krzesło elektryczne.1 }
- synonimia_międzyjęzykowa: Syn_plWN-PWN (1/1)
- hiperonimia_międzyjęzykowa: Hiper_plWN-PWN (4/4)
- hiponimia_międzyjęzykowa: Hipo_plWN-PWN (1/1)

Figure 4-10 Presentation of the noun *krzesło* in Polish WordNet (Wroclaw University of Technology, 2006)

As an example of a terminological ontology, I will consider GOLD, which stands for General Ontology for Linguistic Description. GOLD was developed by Terry Langendoen, Scott Farrar, and William D. Lewis (Whalen, 2004, p. 329). The GOLD ontology gives an account of basic concepts used in linguistics and relations between them. It is based on Crystal's (2008) glossary of linguistic terms which is an appendix to his *Cambridge Encyclopedia of Language* and it is further expanded in Wiki's way (GOLD, 2010). GOLD is a collaborative application which can be extended and applied to all languages. GOLD is an example of specialised ontology, as it describes specialised concepts from the field of linguistics. The entry for interjection in GOLD is presented in Figure 4-11.

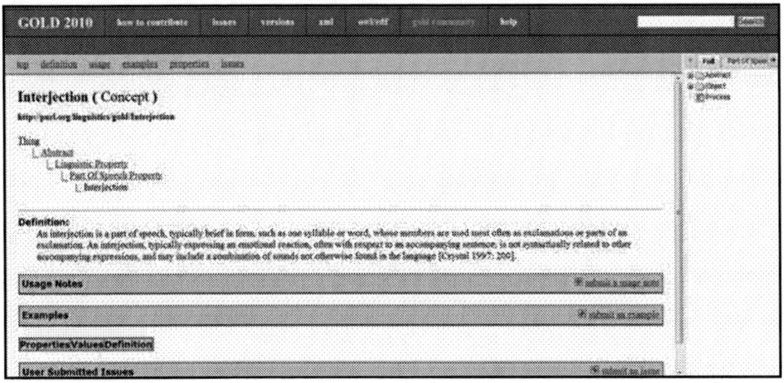

Figure 4-11 The entry for interjection in GOLD (2010)

The GOLD Ontology Viewer is divided into two main panels. The main concept panel on the left side displays information regarding the current GOLD concept. It contains a diagram showing the place of the concept in ontology, the definition of the concept, usage notes, examples and properties. The tree panel on the right hand side displays the full tree hierarchy of the ontology. All immediate child concepts are available upon clicking the plus sign situated to the left of the concept. There are also very useful tree structures, each in their own tab, which include Part of Speech Property, Morphosyntactic Property, Phonetic Property and Morphosemantic Property (GOLD, 2010). GOLD uses data extracted from different sources and illustrates concepts with examples in different languages.

To sum up this discussion of ontologies, it is justified to state that ontologies and termbanks may support each other. Termbanks provide data to instantiate a terminological ontology and the ontology facilitates a very effective data display by means of diagrams which show concepts and relations between them explicitly. Thus, they may be very useful in the analysis of conceptual mismatches.

I looked at off-the-shelf ontologies and tried to find one that would suit the purpose of my research. Off-the-shelf terminological ontologies are very rare. The general ones, by contrast, are common but not very useful for my task. Therefore, I decided to use CAOS to build my own ontology. In the first stage of this process, I examined ontologies as they exist in order to use what I found there for building my own ontology.

It was not possible to build the ontology for all the concepts I collected in my termbase as the whole process of ontology creation is quite demanding and time-consuming. For this reason, I focused on some narrow aspects of surveying, e.g. measuring tools, and tried to build an ontology for this particular sub-domain. I used this ontology in section 4.5 to analyse conceptual mismatches between English and Polish terminology of surveying tools.

4.3 Conceptual mismatches

Conceptual mismatches occur when the concept systems of the source language and the target language differ, resulting in various types of lexical disrepancies. According to Bentivogli and Pianta (2000, p. 664) the most important lexical discrepancies are of two types:

- lexical gaps, which occur when one of the two languages has no lexicalisation for a concept. For example, what English expresses with the lexical unit *aspect of projection*, the Polish language has to express with the free combination of words *położenie płaszczyzny odwzorowawczej względem osii ziemi* 'the situation of the surface of projection with respect to the Earth's axis';
- denotation differences, when the denotation of the target equivalent only partially overlaps with the denotation of the source language word. This situation takes place when the translation equivalent of the source language exists but is more general or more specific. In the former case this translation equivalent is a hyperonym, while in the latter case it is a hyponym of the word in question. For example, the English term *chainman* has an equivalent in Polish in the form of *pomiarowy* 'surveyor's assistant' which has a more general meaning than the English term.

Conceptual mismatches may occur in various areas of surveying, both in technologically-oriented fields and in the legal domain. They occur because concept systems differ between the continental European and the Anglo-Saxon traditions, and also between concepts in American and British English. Concepts vary because countries have their own, individual legal systems and different approaches to understanding the world. Many concepts are culture-specific and do not have corresponding concepts in other languages. Some other concepts have similar counterparts in other languages, but there are still differences which need to be highlighted between them.

I will carry out an analysis of conceptual mismatches by combining the descriptive and normative approaches. For a translator it is important to look at surveying concepts in the source texts (STs) and try to interpret them by looking at how they are actually used. In producing the target text, the translator will observe standards as norms, which indicate how concepts should be used. I will look at concept systems in English and Polish and analyse how they correspond and how they are represented. I will conduct a detailed analysis using data from the English termbase. I will look at the definitions of concepts and the semantic relations they have with other concepts. I will then compare English concepts with Polish concepts by looking at concept names (their literal translations into English as well as standardised equivalents), English translations of their definitions and the semantic relations they have with other Polish concepts.

I will start the analysis of conceptual mismatches in surveying with the technological domain. I will rely in my analysis on the ISO 9849:2000 standard *Optics and optical instruments – Geodetic and surveying instruments – Vocabulary* (ISO, 2000b), *The Glossary of Mapping Sciences* (1994) edited by DeLoach, the Ordnance Survey glossary (2011) compiled by Lawrence and various surveying textbooks. It should be emphasised that the majority of sources in English, including the *Glossary* being the most complete source of surveying terminology in English, were published in the U.S. For Polish, I will use *Polska norma PN-N-02207:1986 Geodezja. Terminologia* 'Polish standard on geodetic terminology' (Polski Komitet Normalizacyjny, 1986) and Polish surveying textbooks.

4.3.1 Case study 1: theodolite vs transit

The first example will consider similarities and differences between the concepts of theodolite and transit, which are presented in Table 4-2.

Table 4-2 Transit vs theodolite

term	definition	synonym
theodolite	instrument for measuring horizontal directions or horizontal directions and vertical angles, whose main components are the horizontal circle and the vertical circle, the telescope and additional devices for reading graduated circles and for setting up the vertical axis ISO 9849:2000	US transit ISO 9849:2000

transit	an astronomical or surveying instrument consisting primarily of a telescope mounted so that it can be rotated about a horizontal axis to describe an arc of about 180° from horizon to horizon DeLoach (1994)	transit theodolite DeLoach (1994)
TRANSIT	the first generation satellite positioning system designed for navy purposes, which operated on the Doppler principle Ghilani & Wolf (2008, p. 323)	

Transit is a problematic term. It is polysemic with two basic senses in surveying: *transit* as a surveying instrument and *TRANSIT* as the satellite positioning system. There is also a third sense: *transit* as an astronomical instrument used to observe the passage (transit) of stars across any portion of the celestial meridian (Duggal, 2009, p. 113). This sense is historically the first sense of the term. It dates back to the 17th century when the instrument was developed, and is hardly used nowadays.

The current usage of the term in surveying is very ambiguous. The term *transit* is sometimes used as a synonym for *theodolite*, e.g. "A transit or theodolite is an instrument used to measure horizontal and vertical angles" (Duggal, 2009, p. 113); as a type of theodolite, i.e. transit theodolite, when its telescope can be revolved through 180° in a vertical plane about its horizontal axis (Roy, 2010, p. 180); or to mean a surveying instrument similar to a theodolite but of insufficient accuracy to permit it to be used for establishing geodetic control (Ghilani & Wolf, 2008, p. 862). DeLoach (1994) claims that the first usage is definitely undesirable, because some theodolites do not permit the telescope to be rotated through 180° vertically. The second usage is undesirable for the same reason and because some instruments referred to as theodolites are not sufficiently accurate for geodetic purposes. To understand this ambiguity, I have to focus on the origin of the instrument and study its development. The first transit in the sense of instrument with a reversing telescope was produced by William Young in Philadelphia in the USA in 1831. The term *transit* was adopted for the instrument because its telescope could be transited or reversed by rotating it about a horizontal axis. In Europe, the name *transiting theodolite* was adopted for this type of instrument. However, Europeans dropped the adjective and retained the name *theodolite* (Ghilani & Wolf, 2008, p. 861). Although in the past theodolites could be divided into transit and non-transit types, such a division is not justified nowadays, as all recently manufactured theodolites are transit theodolites (Roy, 2010, p. 180).

Thus, it may be assumed that *transit*, *transit theodolite* and *theodolite* were synonyms at some point in the history. However, it seems that the design of theodolites has developed, while transits have remained the same, which has resulted in the emergence of two quite different instruments with different sets of features. Ghilani and Wolf (2008, p. 862) argue that there is no internationally accepted understanding among surveyors on the exact difference between the terms *transit* and *theodolite*. The most widely used criterion is their general design, especially their graduated circles and the systems for reading them. Transits have an "open-circle" design which allows an operator to see their graduated circles and read them with the aid of verniers (Figure 4-12).

Figure 4-12 Transit (Photograph by Ewelina Kwiatek, with permission from Faculty of Mining Surveying and Environmental Engineering, AGH Krakow, 2012)

Theodolites, on the other hand, feature an enclosed design (Figure 4-13). They have graduated circles made of glass which are not directly visible to an operator and must be read by means of an internal microscopic optical system. Angle observations performed using theodolites are in general more precise than the readings from transits (Ghilani & Wolf, 2008, p. 863).

Figure 4-13 Theodolite (Photograph by Ewelina Kwiatek, with permission from Faculty of Environmental Engineering and Land Surveying, University of Agriculture in Krakow, 2012)

There is certainly evidence indicating that transit and theodolite are not synonymous. Thus, the interpretation of the concept provided in ISO 9849:2000, which states that *transit* is an American synonym of *theodolite*, oversimplifies the classification and was only true at some earlier point in the history of the development of the two instruments.

I will now look at how the concepts of theodolite and transit are discussed in Polish. The Polish concept system for theodolites is presented in Table 4-3.

Table 4-3 Theodolite in Polish

term	definition
teodolit 'theodolite'	instrument used for horizontal and vertical angle measurement Wołłodko (1973, p. 240)
teodolit mechaniczny 'mechanical theodolite'	theodolite with a metal limb and a vernier for circle readings, featuring low reading accuracy between ±20'' and ±1', used until the1960s Jagielski (2005, p. 114)
teodolit optyczny 'optical theodolite'	theodolite equipped with a glass limb in the shape of a ring, whose diameter is 50-100mm and utilising an optical circle reading through a micrometer Jagielski (2005, p. 115)
teodolit elektroniczny 'electronic theodolite'	theodolite in which the limb is replaced with coded circular plates, which facilitate electronic circle scanning and provide results in a binary format Jagielski (2005, p. 115)
teodolit z układem jednoosiowym 'theodolite with a system of a single-axis'	theodolite in which the limb is permanently coupled with the tribarach and which allows only alidade movements Jagielski (2005, p. 116)
teodolit repetycyjny 'repetition theodolite'	theodolite equipped with a double-vertical axis and a repetition clamp to couple limb and alidade (and to measure horizontal angles a number of times) Jagielski (2005, p. 116)
teodolit reiteracyjny 'reiteration theodolite'	theodolite equipped with a reiteration mechanism used to rotate the limb independently from the alidade Jagielski (2005, p. 116)

The concept of 'transit theodolite' was not encountered in Polish. However, the concept of revolving the telescope through 180° in a vertical plane exists in Polish and is known as *przechylenie* or *przerzucenie lunety przez zenit* 'leaning or shifting the telescope about 180° or through zenith'. The analysis of the type of theodolites in the Polish concept system led to an interesting observation: Polish theodolites are divided into mechanical, optical and electronic on the basis of the type of the circle they have and the method of circle reading. The same classification scheme may be applied to English and identical concepts are recognised. The second criterion is connected with the vertical systems in theodolites. It enables theodolites to be divided into two groups: theodolites with a single vertical axis and theodolites with a double-vertical axis. The latter may be further subdivided into repeating theodolites, in which the limb and alidade can be coupled by means of a clamp and which facilitate the repetition of horizontal angles, and reiteration theodolites, in which the circle may be rotated independently of the alidade (Jagielski, 2005, p. 116).

In English, the concept of 'reiterating' theodolite is hardly known. I found very few cases of the use of this concept. There is a reference to a reiterating theodolite in Carter's thesis (1965), where it is described as an instrument in which the horizontal circle can be displaced, in either direction by any desired amount, by using the circle orienting gear. Theodolites in the English concept system are classified into two types taking the type of vertical axis as a criterion (Ghilani & Wolf, 2008, p. 871):

- repeating theodolites;
- directional theodolites.

The concept of a repeating theodolite is the same in English and Polish. However, the concept of a directional theodolite needs verification, which is possible by studying its definition. A directional theodolite is defined as a theodolite with a single vertical axis, not equipped with a lower motion, which facilitates reading directions rather than angles and has a horizontal circle positioning drive (Roy, 2010, p. 211). If I compare this definition to the definition of the single-axis theodolite in the Polish concept system (Jagielski, 2005, p. 116), I notice that these concepts do match. I will come back to this issue in section 4.5.1.

4.3.2 Case study 2: level vs spirit level

The next problematic area where conceptual mismatches occur relates to the concept of level, which represents a case of polysemy as level may designate both a surveying instrument (holonym) and one of its parts (meronym). Different types of levelling instruments are presented in Table 4-4.

Table 4-4 Levelling instruments and devices

term	definition	synonym
level	instrument for measuring differences in height by establishing horizontal lines of sight, comprising as main components a telescope which can be rotated on a vertical axis and a facility for levelling the line of sight ISO 9849:2000	
spirit level	level with the line of sight levelled by a tubular $level_2$[2] ISO 9849:2000	tilting level, bubble level ISO 9849:2000
compensator level	level with the line of sight automatically levelled by means of an inclination compensator ISO 9849:2000	self-levelling level, pendulum level, automatic level (depre-cated)
digital levei	level for automatic levelling, with CCD sensor system requiring the use of a spherical digital staff, bearing a bar-coded scale ISO 9849:2000	digital electronic level Ghilani and Wolf (2008, p. 91)
dumpy level	level in which the telescope is permanently attached to the base carrying the spirit levels, either rigidly or by a hinge about which the telescope can be rotated by means of a micrometer screw piece (DeLoach, 1994)	
tilting level	levelling instrument in which the line of sight is brought into its final, level position by rotating the telescope on its trunnions DeLoach (1994)	
wye level	level whose telescope rests in supports on the level bar called wyes Ghilani and Wolf (2008, p. 861)	

[2] $level_2$ refers to a part of levelling instrument

The term *level* is a hyperonym for the different types of levels discussed in Table 4-4. These types of levels were not derived from one classification scheme, but are the outcome of a general search for various types of levels and belong to different classification systems. Looking at various classification systems for levelling instruments allows us to establish relations between level types.

ISO 9849:2000 specifies the following types of levels: compensator level (automatic level), digital level, electronic level (instrument which indicates electronically the horizontal position of an object), hydrostatic level (instrument used to determine height differences over great distances or where the height differences are very small and high accuracy is essential), spirit level. This classification is inconsistent with other classification schemes, such as those found in textbooks. Ghilani and Wolf (2008, p. 85) classify levelling instruments into four categories: dumpy levels, tilting levels, automatic levels and digital levels. They claim that a digital level is necessarily electronic, since its readings are in electronic format (2008, p. 91). Their classification also indicates that automatic level is synonymous with compensator level from ISO 9849:2000. Ghilani and Wolf (2008, p. 931) also specify *wye level* as a type of level and describe it as an old instrument that is similar in many respects to the dumpy level. Hydrostatic level, listed in ISO classification, is not mentioned in other sources. This may be because its body does not resemble other levels. It consists of tubes that are connected with a hose, and its working principle is based on the theorem that the calm surface of a liquid in connected tubes forms a horizontal plane (Deumlich, 1982, p. 162).

Levels may be classified on the basis of the components they use to orient their lines of sight horizontally. Spirit levels use vials to orient their lines of sight and they use a levelling staff. Automatic levels employ automatic compensators. Digital levels also use automatic compensators, but use a bar-coded rod for automated digital readings (Roy, 2010, p. 81).

Taking into consideration the way the telescope is mounted, two types of levels may be distinguished: wye level (Y-level) and dumpy level (Duggal, 2009, p. 236). In a wye-level the telescope is supported by a pair of wye rings (metal parts in the shape of the letter Y, hence the name of the instrument), which can be opened for the purpose of turning the telescope or rotating it around its horizontal axis. In the dumpy level, the telescope is permanently attached to the base carrying the spirit levels, either rigidly or by a hinge about which the telescope can be rotated by means of a micrometer screw piece. The tilting level is a variation on the dumpy level. Its telescope can be effectively flipped through 180°, without rotating the head.

The analysis of semantic relations between different types of levels proves that ISO 9849:2000, which is one of the main sources of information in this study, is inconsistent. According to the standard *spirit level* is synonymous with *tilting level.* However, the definitions provided for the two entries do not confirm this statement. Spirit level is defined as a level with the line of sight levelled by a tubular level, while in a tilting level, the line of sight is brought into its final, level position by rotating the telescope on its trunnions. The two definitions are not related in any way. An expert in the field knows that the tilting level has levelling vials as its components and is synonymous with the spirit level. However, a translator with no background in surveying would assume that spirit level and tilting level are different concepts especially if he/she does not find information that they are synonymous.

Another shortcoming of ISO 9849:2000 is reflected in the status of the names. The standard claims that the term *automatic level* is deprecated and *compensator level* should be used instead. However, *automatic level* is used widely in the literature on the subject. Authors of various textbooks, e.g. Ghilani and Wolf (2008), Duggal (2009), Bannister et al. (1998), refer to *automatic levels* rather than *compensator levels*. The ISO standard, which is supposed to guide professionals in the usage of terms, does not seem to have much authority in this case.

The concept of *level* as such is very ambiguous because it is a hyperonym for different types of levels and a meronym for part of a levelling instrument. The concepts related to levelling instrument parts are presented in Table 4-5.

Table 4-5 Parts of levelling instruments

term	photograph	definition	synonym
level		closed hollow vial which is partially filled with liquid, the remaining space containing air which finds its way to the highest point in the vial ISO 9849:2000	spirit level ISO 9849:2000 level vial Ghilani and Wolf (2008: 87)
circular level	 (Photograph by Ewelina Kwiatek, with permission from Faculty of Mining Surveying and Environmental Engineering, AGH Krakow, 2012)	level having the inside surface of its upper part ground to spherical shape ISO 9849:2000	bull's eye level, box bubble, circular bubble ISO 9849:2000 universal level, spherical level DeLoach (1994)
tubular level	 (Photograph by Ewelina Kwiatek, with permission from Faculty of Environmental Engineering and Land Surveying, University of Agriculture, Krakow, Poland, 2012)	level with a tubular glass vial which is barrel-shaped internally and graduated on its upper surface, fixed into a metal holder and fitted with adjusting screws ISO 9849:2000	

The information collected in Table 4-5 indicates that *level* is not only a superordinate concept for different types of levels and a meronym referring to part of a levelling instrument, but also plays the role of hyperonym for different kinds of levelling vials, viz. tubular level and circular level. What is more, it is synonymous with *spirit level*, which is a hyponym of *level* (levelling instrument). Thus, both *level* and *spirit level* are instruments and instrument parts. There is no practical problem arising from this ambiguity in monolingual terminological work and therefore no normative solutions are proposed.

Another type of level which occurs in levelling terminology is *Abney level* (invented by William de Wiveleslie Abney). Although it is considered to be a type of level, it does not have much in common with the types of levels described above and therefore it was not presented along with them in Table 4-4.

An Abney level is a hand-held device consisting of a fixed sighting tube, a movable spirit level that is connected to a pointing arm, and a protractor scale. The Abney level is commonly used by foresters to measure tree height and the steepness of hills.

The range of Polish concepts for levelling instruments is presented in Table 4-6. The Polish semantic network for levelling instruments consists of *niwelator* 'level' as a hyperonym and *niwelator cyfrowy* 'digital level', *niwelator optyczny* 'optical level' and *niwelator libellowy* 'bubble tube level', i.e. spirit level, as the hyponyms. The name of the level type comes from relational adjective that describes the relevant property of the instrument, e.g. the name of *niwelator libellowy* 'bubble tube level' comes from the part of the instrument – *libella* 'the bubble tube', i.e. spirit level, which is used to level the axis of sighting.

The concept system for levelling instrument parts presented in Table 4-7 is similar to the English concept system with *libella* 'level vial' as the hyperonym and *libella pudełkowa* 'box bubble' and *libella rurkowa* 'tubular level' as hyponyms. It is important to highlight that libella has alternative spelling as it may be written down as libela. The former spelling is the one that is prescribed by dictionaries, e.g. Dubisz (2003), whereas the latter is commonly applied by surveyors and may be encountered in text books, e.g. Wołłodko (1973).

Table 4-6 Levelling instruments in Polish

term	definition	synonym
niwelator 'level'	geodetic instrument that is used for height difference measurement Encyklopedia PWN (2010), retrieved on 11.08.2010	
niwelator automatyczny 'automatic level'	levelling instrument, whose axis of sighting is levelled by means of a so called compensatory device Wołłodko (1973, p. 126)	niwelator samopoziomujący 'self-levelling level'
niwelator cyfrowy 'digital level'	levelling instrument that allows for the presentation of staff's reading and distance measurement on a screen and that can be registered in the internal memory of the instrument or in the memory of an external registering device Jagielski (2005, p. 165)	
niwelator libellowy 'spirit level'	levelling instrument, whose axis of sighting is levelled by means of a level Wołłodko (1973, p. 126)	
niwelator optyczny 'optical level'	levelling instrument that is equipped with an optical micrometer Jagielski (2005, p. 161)	

Table 4-7 Levelling instruments' parts in Polish

term	definition	synonym
libella 'level vial'	device used for horizontal or vertical aberration measurement and for the measurement of very small angles by measuring displacement of a gas bubble that is contained in an ampoule with liquid from its zero position Encyklopedia PWN (2010), retrieved on 8.02.2010	
libella pudełkowa 'box bubble'	cylinder-shaped spirit level, the top of which is closed with a spherical cap Wołłodko (1973, p. 126)	libella sferyczna 'spherical spirit level'
libella rurkowa 'tubular level'	spirit level of an oblong cross section in the shape of circle's arc Pasławski (2006b, p. 61)	

The Polish concept system for levelling instruments and their parts is more transparent, as there are not so many synonyms as in English, and therefore there are fewer ambiguities. Levelling instruments and their parts have different names. The levelling instruments are called *niwelatory* 'levels', while the levelling instruments' parts are called *libelle* 'level vials'.

The term *libella* is also used in the construction industry where it is synonymous to *waserwaga* which refers to a wooden or a plastic staff with a built-in cylinder filled with alcohol or ether in which a gas bubble is created (Bańko, 2005). A position of this bubble indicates displacement from the level. Waserwaga is used in brickworks. It is used interchangeably with *poziomica* and *poziomnica.* It comes from German *Wasserwage* (from *Wasser* 'water' and *Wage* 'weight').

4.3.3 Case study 3: surveying vs geodesy

Conceptual mismatches also occur in the name of the field of surveying, which was already briefly discussed in 1.2.1. There is no common international agreement regarding the field name. There are different names and naming conventions in Europe and the USA. Concepts related to the name of the field are presented in Table 4-8.

Table 4-8 Names of the field in English

term	definition	synonym	hyperonym
surveying	the science, art and technology of determining the relative positions of points above, on, or beneath the Earth' s surface, or of establishing such points, and the presentation of this information either graphically or numerically Ghilani and Wolf (2008, p. 1)		
geomatics	the science and technology of gathering, analysing, interpreting, distributing and using geographic information Canada Centre for Remote Sensing (2005), retrieved on 18.02.2011		

geodesy	the science of measuring and monitoring the size and shape of the Earth and the location of points on its surface National Geodetic Service (2010), retrieved on 13.10.2010		
geodetic surveying	branch of surveying in which large areas of the Earth's surface are involved and the curvature of the Earth is taken into account Bannister et al. (1998, p. 1)		surveying
plane surveying	branch of surveying in which relatively small areas are under consideration, and it is assumed that the Earth's surface is flat Bannister et al. (1998, p. 1)		surveying
civil engineer-ing	the profession of designing and executing structural works that serve the general public Encyclopaedia Britannica (2009)		engineering
land surveying	the process of determining boundaries and areas of tracts of land DeLoach (1994)	boundary surveying, cadastral surveying DeLoach (1994)	surveying
boundary surveying	the process of establishing or reestablishing a boundary line on the ground, or of aiming to obtain data for constructing a map or of plotting showing a boundary line DeLoach (1994)		surveying
cadastral surveying	surveying related to land boundaries and subdivisions, whose aim is to create units suitable for transfer or to define the limitations of title Bureau of Land Management (2009), retrieved on 26.04.2009		surveying

It is not very obvious which of the names included in Table 4-8 should be used to name the science and technology of measuring the relative positions of natural and man-made features on the Earth's surface as well as presenting the collected data. In continental European countries, the term *geodesy* is used to name this field but it is rarely applied in Anglo-Saxon countries, where *surveying* seems to be the preferred term. The concept of *geodesy* has been recorded in various textbooks (Bannister et al; Uren & Price, 2010) and is understood as the study of the size and shape of the Earth and its gravity field. It gives rise to the name *geodetic surveying*, which is a branch of surveying, concerned with measurements of large areas of land, such as a whole country. It requires the highest possible standard of measurements which can be achieved by using modern techniques such as satellite positioning systems. The other type of surveying, *plane surveying*, treats the surface of the Earth as if it were flat and deals with measurements of small areas of land.

The concept of surveying has changed a lot over time, which is well-reflected in the discussion of geodetic surveying – nowadays involving satellite positioning techniques. The traditional role of a surveyor was to determine the position of features in natural and built environments on or below the surface of the Earth and to represent it on a map (Ghilani & Wolf, 2008, p. 3). The current role of the surveyor is wider. It requires knowledge of different methods for collecting spatial data, for processing this data in various formats and for presenting this in an assortment of media. Because the emphasis in surveying has shifted from measurements to data processing, and advances have been made in instrumentation for data collection and processing, various organisations such as the Royal Institution of Chartered Surveyors (RICS) or the Chartered Institution of Civil Engineering Surveyors (ICES) argue that a change of name from surveying to geomatics better reflects the nature of the profession as it is today (Uren & Price, 2010, p. 5).

The name *geomatics* has gained widespread acceptance in the United States as well as in other English-speaking countries of the world, especially in Canada, the United Kingdom and Australia. Many college and university programs in the United States that were formerly identified as 'Surveying' or 'Surveying Engineering' are now called 'Geomatics' or 'Geomatic Engineering' (Ghilani & Wolf, 2008, p. 931).

The approach was quite different in Europe, where surveying was used to describe traditional methods of measurements, such as tacheometry, levelling, angle measurement, whereas geomatics referred to the three- and four-dimensional measurement science including image interpretation and sensor network; spatial data infrastructure; and the communication of geographic information including mapping, visualisation and verbalisation. However, quite recently, European countries have been adopting the same conventions as Anglo-Saxon countries and offer courses in Geomatics (AGH University of Science and Technology, 2011).Table also includes the concept of *civil engineering* which does not show any apparent relation with the concept of *surveying*. The term was first used in the 18th century to distinguish the newly recognised profession from military engineering (Encyclopaedia Britannica, 2009).

Civil engineering is traditionally divided into several sub-disciplines including surveying. The majority of engineers working on civil engineering projects use traditional techniques of surveying such as measurement of angles, distances and heights. This gives rise to the term of *engineering surveying*, which is hyponym of surveying, defined as any survey work carried out in connection with construction and building (Uren & Price, 2010, p. 2). *Engineering surveying* is distinguished from *geospatial engineering* which reflects changes in the way in which survey data is collected and processed for civil engineering projects today, and is usually associated with the involvement of photogrammetry, remote sensing, geographic information systems, cartography and visualisation. Thus, the role of a civil engineer involves the tasks performed by a surveyor or a geomatic engineer. His knowledge of measurement methods is usually limited to those currently applied to, and most useful for, the purpose of civil engineering projects.

Apart from *geodesy*, *surveying* and *geomatics* the concept of *land surveying* often occurs in the literature on the subject. DeLoach (1994) defines it as the process of determining boundaries and areas of tracts of land and claims that it is synonymous with boundary surveying and cadastral surveying. However, an in-depth analysis of these three concepts indicates a few differences in their meaning.

DeLoach (1994) argues that the term *boundary survey* is usually restricted to surveys of boundary lines between political territories. For the survey of boundary lines between privately owned parcels of land, the term *land survey is* preferred (although *property survey* is also used); for official surveys of the public lands of the United States of America, *cadastral survey* is used. This specification is, however, in contrast with the guidelines of the Maine Society of Land Surveyors (n.d), which state that a boundary survey determines the property lines of a parcel of land described in a deed. It will also indicate the extent of any easements or encroachments and may show the limitations imposed on the property by state or local regulations.

The definition of *cadastral survey* by DeLoach (1994) is compatible with other definitions. Bannister et al. (1998, p. 3) describes cadastral surveys as surveys that are undertaken to produce plans of property boundaries for legal purposes. In many countries the registration of ownership of land is based on such plans.

Taking all these facts into account, *land survey* seems to be an umbrella concept including both boundary surveys, whose purpose is to establish boundary lines between parcels of land (both political territories and privately owned land) and cadastral surveys which focus on showing ownership of land (through boundary lines).

Boundary surveys, which date back to about 1400 B.C. in Egypt, are considered the oldest types of survey in recorded history (Ghilani & Wolf, 2008, p. 621). Thus, boundary surveys along with land surveys were the predecessors of geodesy, surveying and geomatics. The terms *boundary surveys* and *land surveys* are still in use, but nowadays they delineate the type of survey carried out as a part of a larger project rather than naming the field.

The concept system for Polish names of the domain in question is quite different. It is presented in Table 4-9.

Table 4-9 Concepts behind the field names in Polish

term	definition	synonym	hyperonym
geodezja 'land surveying'	Earth science, the aim of which is to determine the Earth's shape and sizes, prepare the mathematical model of the real Earth's globe, as well as to determine mutual position of points situated on the Earth's surface Jagielski (2005, p. 10)	surveying, geodesy	
geodezja ogólna 'general surveying', i.e. plane surveying	branch of surveying that deals with measurements and preparation of large-scale maps of small areas that can be referred to a surface without taking into consideration the Earth's curvature Jagielski (2005, p. 11)	miernictwo geodezyjneo r geodezja niższa	surveying

geodezja wyższa 'superior surveying', i.e. geodetic surveying	branch of surveying that deals with studies of the shape and size of the Earth, as well as measurements of larger areas taking into consideration reference to the surface's curvature Jagielski (2005, p. 12)		surveying
geodezja gospoda-rcza 'economic surveying'	branch of surveying that includes a wide range of geodesy methods in such economic fields as administration, industry, communication, agriculture, forestry, mining and railways Jagielski (2005, p. 12)		surveying
geodezja dynami-czna 'dynamic geodesy'	branch of surveying which is concerned with the delineation of the shape and location of a geoid on the basis of gravimetric measurements Jagielski (2005, p. 12)		surveying
geoinfo-rmatyka 'geo-informatics, geomatics'	scientific and technical discipline, dealing with the application of computer science in land-surveying studies in order to retrieve, process, analyse and share geographical information KARTO (2005), retrieved on 15.01.2009		surveying
kartografia 'cartogra-phy'	branch of science that includes the theory and methods of producing and using maps, atlases, globes and models representing the Earth or other celestial bodies Encyklopedia PWN (2010), retrieved on 8.02.2010		surveying

topografia 'topogra-phy'	discipline that deals with the preparation of general geographical maps in the following scales: 1:5,000, 1:10,000, 1:25,000, 1:50,000, 1:100,000, that are made on the basis of large scale studies or a distinct measurement technique Jagielski (2005, p. 12)		surveying
fotograme-tria 'photogram-metry'	branch of surveying, which deals with spatial data acquisition by registration, measurement, processing and interpretation of photograms KARTO (2005), retrieved on 15.01.2009		surveying
instrume-ntozna-wstwo geodezyjne 'knowledge study of surveying equipment'	branch of surveying which deals with the construction, examination, usage and maintenance of surveying instruments Jagielski (2005, p. 12)		surveying
rachunek wyrówna-wczy 'adjustment methods'	branch of surveying which deals with methods of surveying calculations, the adjustment of measurements results (observations) and unknown values in order to delineate their most probable values and evaluate the accuracy of measurement and adjusted values as well as with the optimalisation of surveying works Jagielski (2005, p. 12)		surveying

astronomia geodezyjna 'geodetic astronomy'	branch of surveying which determines the location of points and orientation of bearings on the Earth's surface by observing celestial bodies Jagielski (2005, p. 12)		surveying

The Polish concept system does not seem to have many common features with the English concept system. It does look more transparent when compared to the English concept system as it consists of the field name *geodezja* 'geodesy, surveying', and eleven hyponyms, which are considered sub-domains of surveying. A shared feature of the English and Polish concept systems is the occurrence of plane and geodetic surveying as branches of surveying. Actually, in English these are rather types of surveying, while in Polish these concepts delineate branches or sub-domains in surveying. *Geodezja* in Polish still has a very strong position as the name of the field and it gives way to *geomatyka* 'geomatics' at a much slower rate than in English. Jagielski (2005, p. 12) uses the term *geomatyka* to refer to retrieving, processing, analysing and sharing geographical information rather than the whole set of surveying tasks which also include taking measurements.

The Polish concept system includes such concepts as *geodezja gospodarcza* 'economic geodesy', *geodezja dynamiczna* 'dynamic geodesy' or *instrumentoznawstwo geodezyjne* 'geodetic instruments', which do not occur in English. It also covers disciplines such as *topografia* 'topography' and *astronomia geodezyjna* 'geodetic astronomy', which in English are considered separate fields of knowledge and are not perceived as sub-domains of surveying.

4.3.4 Case study 4: surveying assistant vs chainman

In Polish, there is a single concept of *pomiarowy* 'measuring man', i.e. surveying assistant, who supports the surveyor in his work by performing a wide range of duties, e.g. holding a chain, shading the surveyor with an umbrella, making readings with instrument.

In English, a distinction is made between different types of surveying assistant roles, which is reflected in different names. In Table 4-10, I collected various concepts related to a surveying assistant.

Table 4-10 Surveying assistant vs chainman

term	definition
surveying assistant	
chainman	an assistant in the process of measuring the length of a line with a chain Bannister et al. (1998, p. 18)
staff person/rod person	an assistant in the process of levelling whose task is to hold a level rod plumb on the correct monument or turning point to give the correct reading Ghilani and Wolf (2008, p. 105)
tape person	an assistant in the process of taping the length on the ground, who holds either the beginning or the end of the tape Ghilani and Wolf (2008, p. 127)
instrumentman	an assistant who makes readings using the instrument OS Project Software (2010)
umbrellaman	an assistant who holds the umbrella over the instrument in order to shade it OS Project Software (2010)

4.3.5 Case study 5: aspect of projection, tangency and case

Conceptual mismatches also occur in the field of cartography. They may be found in the concept system of map projections, particularly in the criteria of dividing map projections into different types.

Map projections may be divided into three basic classes: conic projections (the Earth is projected onto a cone), cylindrical projections (the Earth is projected onto a cylinder) and planar or azimuthal projections (the Earth is projected onto a plane). The criteria for such a division are the aspect of projection and the tangency, presented in Table 4-11.

Table 4-11 Aspect of projection vs tangency

term	definition	synonym
aspect of projection	the orientation or position of a map projection graticule in respect to the lines of latitude and longitude on the globe they are representing The Atlas of Canada (2009)	case DeLoach (1994) The Atlas of Canada (2009)
tangency	the location or locations where a projection surface touches or cuts through the globe Garo (2000)	case Garo (2000)

The definitions of the *aspect of projection* and *tangency* suggest that these are two very different concepts. It is quite strange, however, that they have the same synonym. I will look at how projections can be classified taking aspect and tangency as the subdivision criteria.

According to the Institute of Discrete Mathematics and Geometry, Vienna University of Technology (2008) *aspect of projection* describes the mutual position of the axis of the Earth and the axis of the projection. Three types of aspects may be distinguished: normal aspect (the two axes coincide), transverse aspect (the axis of the projection belongs to the plane of the equator) and oblique aspect (axis of the projection is neither normal nor transverse).

DeLoach (1994) specifies the same set of aspects, but on the basis of how they are produced. Thus, the normal aspect of the map projection is the one that produces the simplest graticule (meridians and parallels are represented as straight lines). The transverse aspect is produced by rotating the ellipsoid through 90° from its position in the normal aspect. An oblique aspect is created by any rotation between 0° and 90°.

DeLoach (1994) makes a second distinction between the polar aspect, which occurs when the centre of the graticule represents a pole of the rotational ellipsoid; the meridional or equatorial aspect when the centre of the graticule represents a point on the equator point; or the oblique aspect, when the appearance of the graticule represents neither of these two.

These classification schemes are quite confusing, because applying two different sets of criteria should provide two sets of different aspects, which does not happen as the oblique aspect occurs in both schemata. DeLoach (1994) admits that there is no standard or generally accepted terminology for the concept of 'aspect'.

Garo (2000) merges these two classifications into one, claiming that each projection surface (family) can be positioned over the globe from one of four aspects: polar, equatorial, transverse or oblique. He also perceives *tangency* as a synonymous concept to *case*, which is inconsistent with presentation of the concept in other sources, e.g. *The Atlas of Canada* (2009) or DeLoach (1994). Garo's interpretation of aspect of projection may be very well illustrated with Figure 4-14 taken from *The Atlas of Canada.*

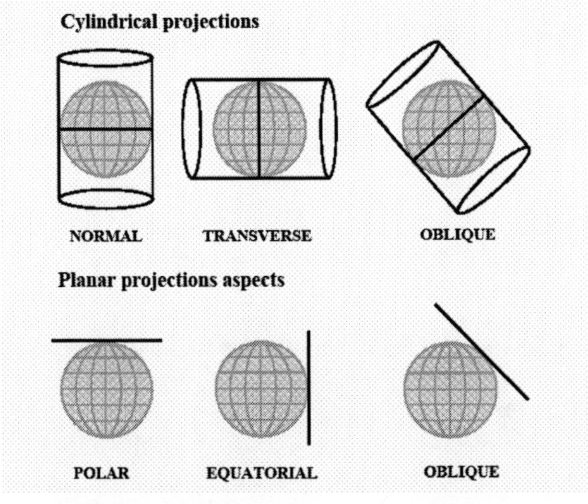

Figure 4-14 Aspects for cylindrical and planar projections (Reproduced with the permission of Natural Resources Canada (2012), courtesy of the Atlas of Canada.)

Figure 4-14 clarifies ambiguities regarding the issue of aspect and allows us to establish a consistent classification scheme for the concept of 'aspect'. Aspect of projection designates how the projection surface is positioned in reference to the globe. Cylindrical and conical projections can be positioned over the globe from the normal, transverse and oblique aspect, while planar projections can be positioned from the polar, equatorial and oblique aspect.

Cylinder, cone and plane can either be tangent or secant to the spheroid. In the tangent case the cone, cylinder or plane just touches the Earth along a single line or at a point. In the secant case, the cone, or cylinder intersects or cuts through the Earth as two circles. The concept of tangency is illustrated in Figure 4-15.

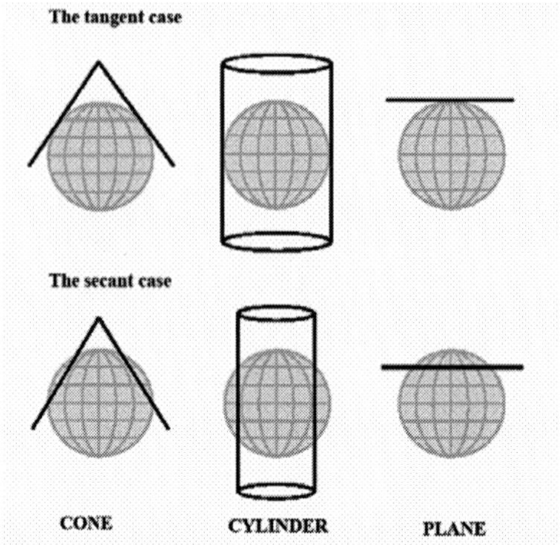

Figure 4-15 Tangency (Reproduced with the permission of Natural Resources Canada (2012), courtesy of the Atlas of Canada.)

On the basis of these facts, I may judge that the concept of *case* refers to two things:

- the map projection graticule in respect of the lines of latitude and longitude on the globe they are representing;
- the location or locations where a projection surface touches or cuts through the globe.

Although the term *case* refers both to the aspect of projection and to the tangency or secancy of the projection, aspect and tangency are in fact completely different features of map projection. The type of projection aspect is used as a projection name, so instead of saying a plane projection with the polar aspect, the name *polar projection* is used to indicate 'an azimuthal projection drawn to show Arctic and Antarctic areas, based on a plane perpendicular to the Earth's axis in contact with the North or South Pole, limited to 10 or 15 degrees from the poles' (Encyclopaedia Britannica, 2009).

Types of projection can be classified on the basis of another criterion. This refers to the point of perspective, which may be the centre of the Earth, a location opposite to the point of tangency, or an external point, as if the Earth were viewed from outer space (The Atlas of Canada, 2009). There is also a concept of perspective projection, which designates the projection defined by a set of straight lines passing through corresponding points on the two surfaces and through a single point common to the set. The common point is called a *perspective centre* (DeLoach, 1994). The perspective refers mainly to a planar family of projections.

To sum up, the English concept system of projection relies on four criteria:

- the type of projection surface (planar or azimuthal projection, cylindrical and conical projections);
- aspect of projection (cylindrical and conical projections are normal, transverse or oblique, while planar projections are polar, equatorial or oblique);
- tangency or secancy (tangent projection, secant projection);
- point of perspective (central projection, stereographic projection, orthographic projection).

In Polish, the concept system of projection types is also built around four criteria (Ogorzelska, 2006, p. 84):

- *rodzaj powierzchni rzutu* 'the type of the surface of projection' onto which a geographical grid representing the sphere/ellipsoid is mapped (planar or azimutal projection, cylindrical projection, conical projection);
- *położenie środka rzutu* 'the location of the centre of projection' from which the sphere is projected on to the surface (central projection, stereographic projection, orthographic projection);
- *położenie powierzchni rzutu w stosunku do bieguna kuli* 'the location of the surface of projection with reference to the pole of the sphere' (cylindrical and conical projections are normal, transverse and oblique, while planar projections are polar, equatorial or oblique);
- *odległość powierzchni rzutu od kuli* 'the distance between the surface of projection and the sphere' (tangent projection is referred to when the projection surface touches the sphere, secant projection when the projection surface cuts the sphere and distant projection, when the projection surface does not touch the sphere and there are no joint points between the two).

The criteria seem to be nearly the same for the English and Polish concept systems. The only difference is the secancy/tangency criterion in the English concept system and the distance between the projection surface and the sphere in the Polish concept system. The Polish criterion is wider and involves the concept of *distant projection*, which is not lexicalised in the English concept system.

English concepts are better designated compared to the Polish concepts. While English employs specific names to refer to different criteria of projection classification, such as secancy/tangency, aspect of projection and point of perspective, Polish does not have such names and relies purely on the description of these concepts.

Polish applies the formulation *rzut* as a synonym for *odwzorowanie*. Although they both mean 'projection', *rzut* has a stronger connotation with cartography as it is associated with projecting (or throwing the image of) the sphere onto the surface, while *odwzorowanie* (from *wzór* 'pattern, design') connotates with 'making an image'.

4.3.6 Case study 6: mapping methods

Conceptual mismatches are also encountered in the classification systems of English and Polish mapping methods. The systems of the two languages vary in many respects and it is often very difficult to find correspondences between concepts in the two systems. When dealing with mapping methods, the English system focuses on the types of maps that are derived by the application of these methods, while the methods as such are often neglected. In many cases, it is difficult to find the precise name of the method that leads to the creation of a particular type of map. On the contrary, the Polish system addresses the methods explicitly.

In my examination of conceptual mismatches in this field, I will first look at the types of thematic maps that are distinguished in the English system and then try to find names for the mapping methods. I will then move to a discussion of the Polish system.

In English, maps are classified as either general-purpose or thematic. *General-purpose maps*, also known as *reference maps*, display natural and man-made objects from the geographical environment, e.g. coastlines, lakes, ponds, rivers, canals, political boundaries, roads, houses. Examples of such maps are topographic maps and road maps. *Thematic maps*, which are also called *special-purpose* or *single-topic maps* are designed to demonstrate particular features, e.g. population, gross domestic product (GDP), etc. Thematic maps may be subdivided into qualitative and quantitative maps. Qualitative thematic maps show the spatial distribution of different forms of geographical phenomena, e.g. distribution of coal fields in the USA. They are used for data that have no magnitude or size difference between classes and which are unordered and show only differences in kind such as land use or vegetation types (Tyner, 2010, p. 66). Quantitative maps, on the other hand, depict the distribution of quantitative amounts, e.g. population in cities in the UK (Dent, 1996, p. 8). Kraak and Ormeling (2003, p. 122) present nine different types of thematic maps which are described in Table 4-12. Synonyms are in brackets in the 'type of map' column.

The classification of map types by Kraak and Ormeling (2003) addresses the whole system of thematic maps and does not specify explicitly which maps are of a qualitative type and which maps belong to the category of quantitative maps. The definition of the chorochromatic map as a map which renders nominal values is a key to establishing that this map is a qualitative map. The list of thematic quantitative maps by Dent (1996, p. 183), which includes choropleth maps, dot maps, proportional point symbol maps, isarithmic (isoline) maps, value-by-area maps, and flow maps, clarifies the quantitative status of most of the maps presented by Kraak and Ormeling (2003). Dent (1996) does not mention diagram maps and statistical surfaces. The quantitative nature of statistical surfaces may be easily established, as their definition states that they are a three-dimensional representation of quantitative data. Diagram maps are also based on quantitative data but their usage is generally not advocated as diagrams presented against the map background cause too many distracting graphical cues. Diagram maps may be used for analytic purposes but they are not suited for communication (Kraak & Ormeling, 2003, p. 138).

Kraak and Ormeling (2003), in their discussion of thematic maps, briefly refer to a subdivision of map types according to Freitag (1992) which is based on measurement scale, corresponding graphical variables and (dis)continuity of the data.

Freitag's classification is much more detailed, with many additional map categories being distinguished, e.g. stepped statistical surface, smooth statistical surface. An interesting feature of this classification, which makes it different from other classification schemes, is the presence of types of qualitative maps that have not been mentioned in other systems. It is noteworthy that English classification systems for thematic maps discuss quantitative maps extensively, but they seem to ignore the category of qualitative maps, which are either only briefly mentioned or not considered in the discussion at all.

Table 4-12 Types of thematic maps

type of map	definition
chorochromatic map (mosaic map) Kraak and Ormeling, (2003, p. 129)	map which renders nominal values for areas through different colours as well as through black and white patterns Kraak and Ormeling (2003, p. 129)
choropleth map (area or shaded map, enumeration map) Dent (1996, p. 123)	map which employs distinctive colour or shading applied to administrative or statistical areas Dent (1996, p. 123)
isoline map	map which employs lines that connect points with an equal value Kraak and Ormeling (2003, p. 133)
diagram map	map which uses diagrams to present a given phenomenon Kraak and Ormeling (2003, p. 137)
dot map	map which represents point data through symbols that each denote the same quantity and that have been located as accurately as possible in the locations where the phenomenon occurs Kraak and Ormeling (2003, p. 139)
flow line map Kraak and Ormeling (2003, p. 140) (flow map, dynamic map) Dent (1996, p. 218)	map showing linear movement between places by line symbols Dent (1996, p. 218)
statistical surface (data model) Kraak and Ormeling (2003, p. 141)	the three-dimensional representation of quantitative data used in choropleth and isoline maps for analytic purposes Kraak and Ormeling (2003, p. 141)
cartogram (value-by-area map) Dent (1996, p. 203)	maps drawn in such a way that the areas of the internal enumeration units are proportional to the data they represent Dent (1996, p. 203)
proportional point symbol map	map in which point data are represented by a symbol whose size varies with the data values Dent (1996, p. 180)

Freitag (1992) lists such qualitative map types as nominal point symbol maps, nominal line symbol maps, R.S. (remote sensing) land use maps and chorochromatic mosaic maps. The category of nominal point/line symbol maps is of particular interest because it seems to be a qualitative equivalent of the quantitative proportional point symbol maps. In fact, Kraak and Ormeling (2003, p. 135) address the category of nominal point data claiming that nominal data valid for point locations are represented by symbols that are different in shape, orientation or colour. Symbols may be divided into figurative and geometrical. Figurative symbols are used when associations with the real object might ease recognition. Geometrical symbols are used for more abstract phenomena and usually require a legend in which their meaning is explained.

The classification system for thematic maps in the Anglo-Saxon tradition, which emerges from the classifications presented above, delineates the following types and subtypes of thematic maps:

A. Qualitative maps:
- Nominal point/line symbol map;
- R.S. land use map (map which shows different forms of land use, e.g. major urban areas, horticulture and arable areas, woodlands and forests, moorlands, etc. using remote sensing data);
- Chorochromatic map.

B. Quantitative maps:
- Choropleth map;
- Isoline map;
- Dot map;
- Diagram map;
- Proportional point symbol map;
- Cartogram map;
- Flow line map;
- Statistical surface map.

The classification of map types by Kraak and Ormeling (2003, p. 122) lacks consistency in addressing mapping methods. The authors sometimes refer to map names, e.g. *choropleth map* and sometimes to methods that are used to produce maps, e.g. *absolute proportional method.* In many cases, drawing the correspondence between the name of the map and the name of the method is difficult and can be accomplished only by consulting other cartographic textbooks, e.g. Cuff and Mattson (1982), Dent (1996). Dent (1996, p. 230) specifies six different mapping techniques:

choropleth mapping, the common dot-mapping method, proportional symbol mapping, isarithmic mapping, value-by-area or cartogram mapping, isarithmic mapping, and flow mapping. Thus, it may be noticed that there is a general tendency to derive names of mapping methods from map names. The complete list of mapping names and relevant mapping techniques is presented in Table 4-13.

Table 4-13 Maps and mapping methods in English

	mapping method	**map**
Quali-tative maps	nominal point/line symbol mapping	nominal point/line symbol map
	land use mapping	land use map
	chorochromatic mapping	chorochromatic map
Quantitative maps	choropleth mapping	choropleth map
	dot mapping	dot map
	isarithmic mapping	isarithmic map
	value-by-area/cartogram mapping	value-by area map or cartogram
	proportional point symbol mapping	proportional point symbol map
	flow mapping	flow map
	diagram mapping	diagram map
	the statistical surfaces method	statistical surfaces

The Polish classification of mapping methods differs from the Anglo-Saxon classification in this respect: it focuses on cartographic representation techniques rather than on the types of maps that result from the application of these techniques. The Polish classification of mapping methods was developed by Ratajski (1989). It takes the level of content presented on the map as the criterion and it makes a basic distinction between qualitative and quantitative methods (Table 4-14).

Table 4-14 Mapping methods in Polish

	method	definition
metody jakościowe 'qualitative methods'	metoda sygnatur 'method of signatures', i.e. symbol method	method which presents objects by means of symbols Kasprzak (2009)
	metoda chorochromatyczna 'chorochromatic method' also called metoda powierzchniowa 'area method'	method which depends on highlighting a particular feature on an entire area that is presented on a map and highlighting – on the basis of this feature – spatial units that are different from the qualitative point of view, used to represent areas which do not overlap Okła (2009) and Pasławski (2006b, p. 205)
	metoda zasięgów 'method of ranges', i.e. spatial reach method	method that consists in marking a map's area, where a given phenomenon occurs, by means of linear, spot, signature and descriptive spatial reach of occurrence, used to represent areas which overlap Okła (2009) and Pasławski (2006b, p. 205)
metody ilościowe 'quantitative methods'	metoda kropkowa 'dot method'	method that consists in marking a given map's area with dots (every dot is ascribed to a certain number of objects) that suggest their real location as accurately as possible Okła (2009)

	metoda kartogramu 'cartogram method'	method presenting the intensity of a given phenomenon within the reference units using colour or shaded symbols Pasławski (2006b, p. 214)
	metoda izolinii 'isoline method'	method which employs lines that connect points with an equal value Pasławski (2006b, p. 220)
	metoda kartodiagramu 'diagram method'	method which presents the size and structure of a phenomenon by means of charts and diagrams Pasławski (2006b, p. 226)

When I compare English and Polish classification systems for mapping methods, I identify similarities and differences. The English system specifies three qualitative map types, and so does the Polish system. The *chorochromatic method* is, without doubt, understood in the same way in both systems. The Polish *metoda sygnatur* 'method of signatures', i.e. symbol method is equivalent to the English *nominal point/line symbol mapping*. There is also a concept *metoda zasięgów* 'method of ranges', i.e. spatial reach method, which seems to be similar to the English concept of *R.S. land use map*, as they both present areas that do not overlap. In fact, the concept *metoda zasięgów* is wider, because it may represent any type of phenomena that do not overlap, e.g. spatial reach of the elk population, while the land use map is limited to rendering the types of land use. Remote sensing techniques are used to map land use, as this phenomenon is static and changes annually if farmers decide to rotate crops. Therefore, it is sufficient to use pictures taken by satellites on one day in a year and draw maps on the basis of them. Showing the spatial reach of the elk would be more problematic using this method as it would involve regular monitoring of the movement of these animals using satellite pictures, which is expensive and provides reliable results only when trees do not have leaves as, only then, animals can be recognised in pictures.

The classification of quantitative methods is even more complex. The Polish system with its four mapping methods seems to be very basic when compared to the English system. However, it is relatively easy to match most of the Polish methods with the English methods, as *metoda kropkowa* 'method of dots', *metoda izolinii* 'method of isolines' and *metoda kartodiagramu* 'method of diagram' are the same as the common dot-mapping method, the isarithmic mapping and diagram mapping. The English system has proportional point symbol mapping, which is not specified as a separate category in the Polish system for mapping methods by Pasławski (2006b). However, it is mentioned that *sygnatury ilościowe* 'signatures of quantity', i.e. proportional point symbols also exist and proportional point symbol mapping is used to show quantitative data on the map, e.g. the size of the city (in terms of the population) is expressed with circles of different diameters.

The category of Polish *kartogram* 'cartogram' and English *cartogram* is certainly the most problematic one from the point of view of the different classification systems, as English *cartogram* and Polish *kartogram* seem to be false friends and cause concept matching problems. The analysis of pictures illustrating *kartogram* in Polish in Pasławski (2006b), *cartogram* and *choropleth map* in English in Dent (1996) allows us to establish that Polish *kartogram* is equivalent to *choropleth map* in English, while the English *cartogram* does not have any representation in the concept system highlighted so far for Polish. I will now look at the definitions and representation of these concepts in more detail to confirm my findings and I will try to find correspondence between the different concepts.

A choropleth map is a thematic map in which areas are coloured or shaded to create darker or lighter areas in proportion to the density of a particular characteristic of the theme subject in that area (The Atlas of Canada, 2009). Choropleth maps show data that is geographically bounded by administrative areas, such as states, counties and townships (Johnson, 2003, p. 45). They show density of population, rate of heat flow through the crust, etc (DeLoach, 1994). They are also referred to as anamorphic maps. On choropleth maps enumeration units, such as states or countries, are shaded a particular colour depending on that unit data value. Cartographic convention assigns the darkest colour to the highest value (Cartisan AGS, 2010). A choropleth map illustrating the population in Albania is presented in Figure 4-16.

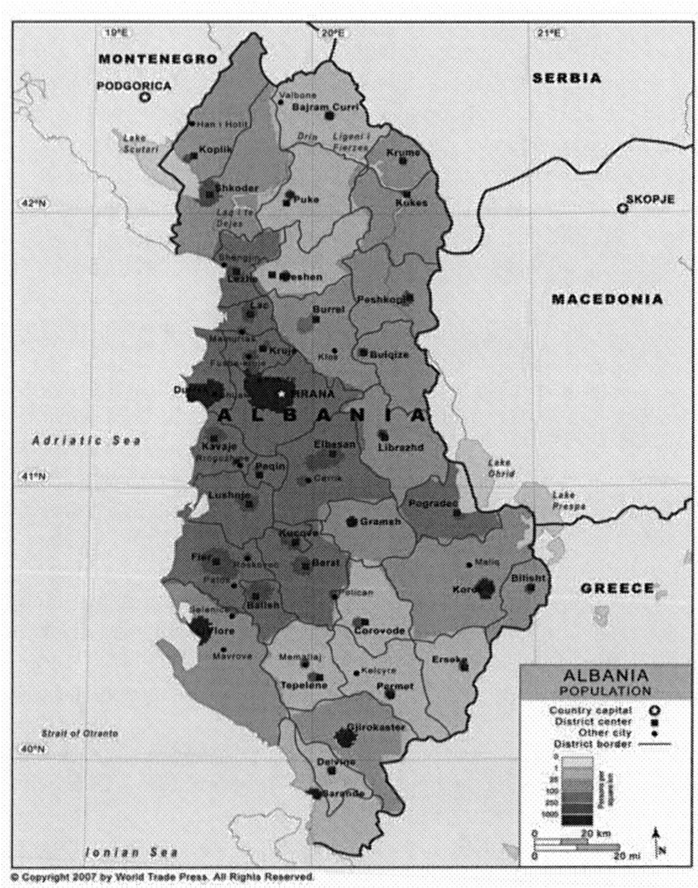

Figure 4-16 Choropleth map of the population of Albania

This choropleth map does not recognise the internal differences in population in individual countries, for example, there is no reference to the fact that people in Canada live mainly along the coastline. When cartographers wish to show the internal distribution of the given phenomenon, they create dasymetric maps, which show the different intensity of the mapped value within the distribution units (Campbell, 2000, p. 175). Dasymetric maps do not rely on enumeration units but combine areas of similar values to depict geographic patterns on the map (Ritter, 2006).

A cartogram is a diagram or abstract map, not to scale, showing quantitative data, by distorting or exaggerating the size of areas (The Atlas of Canada, 2009). A cartogram is also called a *distorted map* or *value-by-area map* (Dent, 1996). The idea of a cartogram may be very well illustrated with a map of world population. While on a traditional world map presenting population data, the sizes of the countries are in proportion to their actual sizes on the surface of the planet and their shapes are the same as their actual shapes (obviously taking into account distortions which result from going from a sphere into a flat map) and only the hue indicates countries with the highest populations, on cartograms the sizes of the countries are proportional to the number of people living there. An example of such a cartogram is given in Figure 4-17.

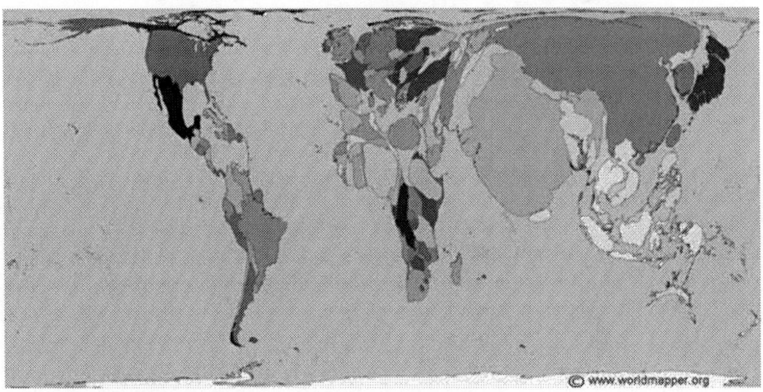

Figure 4-17 Cartogram of the total population of countries in the world. Copyright SASI Group (University of Sheffield) and Mark Newman (University of Michigan). (Worldmapper, 2006)

If I look at the map, I can see that India, China and Japan have become very large. It is due to the fact that the number of people living in these three countries accounts for more than a third of the population of the world. On the other hand, Canada and Russia, the world's two largest countries by land area, have nearly disappeared from the map as they have relatively small populations.

Cartograms are most often used to show population data, but they can be used to show almost any quantity. Figure 4-18 presents a cartogram of the countries of the world in which the sizes of countries are proportional to Gross Domestic Product (GDP). America and Europe dominate this map, while Africa is almost invisible.

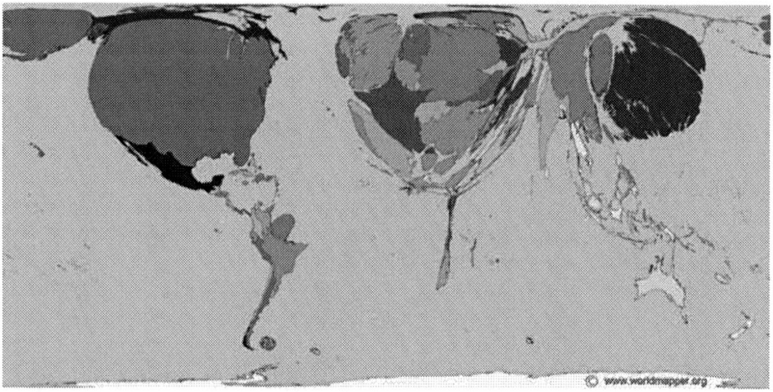

Figure 4-18 Cartogram of the Gross Domestic Product of countries in the world. Copyright SASI Group (University of Sheffield) and Mark Newman (University of Michigan). (Worldmapper, 2006)

Cartograms are often coloured like choropleth maps and therefore these two are often confused. However, in the pictures shown above it is perspective rather than shading that shows the dimension of the value.

The Polish concept system regarding choropleth maps, cartograms and dasymetric maps differs from the English concept system. There is a concept of *kartogram* in Polish, which is synonymous with *choropleth map*, and the concept of *dasymetric cartogram*, which is equivalent to *dasymetric map*. Polish concepts are defined in Table 4-15 below.

Table 4-15 Kartogram in the Polish concept system

term	definition	hyperonym
kartogram 'cartogram'	map presenting the intensity of a given phenomenon within the reference units using colour or shaded symbols Runge and Runge (2009)	mapa 'map'
kartogram dazymetryczny 'dasymetric cartogram', i.e. dasymetric map	a type of cartogram in which the coloured or shaded fields are not enumeration units, but are created to show the intensity of a given phenomenon Pasławski (2006b, p. 214)	kartogram 'choropleth map'

The analysis of the definitions and features of *dasymetric map* and *kartogram dazymetryczny* 'dasymetric cartogram' enables us to establish that the two concepts are equivalent. So does the definition and representation of *choropleth map* in English and *kartogram* in Polish.

The question may arise of how the English concept of *cartogram* can be mapped on to the Polish system. By discussing the equivalency problem with Katarzyna Galant, a Polish researcher working in the field of cartography at Wrocław University of Environmental and Life Sciences, I managed to establish, by comparing definitions and different pictures of English and Polish maps, that Polish cartographers refer to cartograms as *mapy anamorficzne* 'anamorphic maps' or *pseudokartogramy* 'pseudo-cartograms'.

There are two English mapping methods which do not seem to be properly presented in the Polish system. These are the statistical surfaces method and flow mapping. They are not listed as separate methods but they are widely used. Isoline maps may be easily transformed into 3D models which are statistical surfaces. In Polish, the concept of *flow map* as a distinctive type of map is not encountered, but there is *metoda kartodiagramu* 'diagram method'. One of the outcomes of using this method in cartography is the production of *kartodiagram liniowy* 'line diagram'. The line diagrams illustrate connections between areas (e.g. countries) or points (e.g. cities) and their intensity. The width of the line expresses the intensity of the connection. The course of the line may correspond to a rail or road route. A line diagram showing bus connections from Warsaw is presented in Pasławski (2006a, p. 228).

This line diagram is very similar to the flow map found in Kraak & Ormeling (2003, p. 141), which shows transportation achievements and comparison of the actual coal quantities from various areas to Lorraine. This fact confirms that the concept systems of the two languages differ, but scientists work out ways of reaching the same target. In the English concept system, diagrams are a deprecated form of cartographic presentation and probably for this reason flow maps were introduced. In Polish, *kartodiagramy* 'diagram maps' are still listed as one of the main forms of presentation.

4.3.7 Case study 7: system of route classification in the UK and Poland

The United Kingdom and Poland have very different systems for the classification of routes over which the public have right of way. The British system relies on legislation, while the Polish one is more oriented towards tourism and does not have a strong legal basis. It is worth mentioning that when speaking about British rights of way I refer to the rights of way in England and Wales only. Scottish laws are less extensive as Scotland has a long tradition of access to the land (Ramblers, 2011). It is important to note that, in the British land law, public rights of way are often referred to as highways. The concept system of public rights of way in England and Wales is presented in Table 4-16.

Table 4-16 Concept system of the public rights of way in the UK

term	definition	synonym	hyperonym
highway	public right of way over a defined linear route Sydenham (2001, p. 1)	public right of way Sydenham (2001, p. 1)	
byway open to all traffic	highway along which the public have a right of way for vehicular and all other kinds of traffic, but which is used by the public mainly for the purposes for which footpaths and bridleways are so used Wildlife and Countryside Act 1981		highway
bridleway	highway over which the public have the following, but no other, rights of way, that is to say, a right of way on foot and a right of way on horseback or leading a horse and the right to ride a bicycle Road Traffic Act 1988		highway
public footpath (footpath)	highway over which the public has a right of way on foot Sydenham (2001, p. 4)		highway

road used as a public path (RUPP)	highway, other than a public path, used by the public mainly for the purpose for which footpaths and bridleways are used Sydenham (2001, p. 4)		highway
restricted byway	highway over which the public have the following rights: a right of way on foot, on horseback or leading a horse and a right to use non-motorised vehicles including horse-drawn carts Sydenham (2001, p. 4)		highway

The concept system of public rights of way relies on the following legal acts:

- Countryside and Rights of Way Act 2000 (CROW Act 2000);
- Countryside Act 1968;
- Highways Act 1980.

In the reconstruction of this concept system, I also consulted BS 766-4:2002 "Spatial data-sets for geographical referencing - Part 4: Specification for recording public rights of way" (British Standard Institution, 2002) and a textbook by Sydenham (2001). Legal acts have the highest legal authority but they need to be read together as the later acts update information included in the earlier acts. Moreover, they are not focused on providing terminological definitions. British Standard BS 766-4:2002 is the most recent source of information, but it does not relate concepts very well. The textbook by Sydenham is just one year older but it provides the most useful terminological definitions and relates concepts well by showing hyponymy and synonymy relations. It also presents a historical overview of concepts highlighting how they have changed. Therefore, the textbook was the main source of data in the reconstruction of the concept system for rights of way in the UK.

The term *highway*, which is used in general language as a synonym of *motorway*, in the context of rights of way is used interchangeably with the term *public right of way* and is interpreted as a path that anyone has the legal right to use on foot and sometimes using other modes of transport. There are four main categories of highways (Ramblers, 2011):

- public footpaths, which are open only to walkers;
- public bridleways, which are open to walkers, horse-riders and pedal cyclists;
- byways open to all traffic (BOATs), which are open to all classes of traffic, including motor vehicles. However, they may not be maintained to the same standard as ordinary roads;
- restricted byways, which are open to walkers, horse-riders, and drivers/riders of non-mechanically propelled vehicles such as horse-drawn carriages and pedal cycles.

There is still a category of a 'road used as a public path' (RUPP), which was introduced by the National Parks and Access to the Countryside Act 1949. RUPP was quite a vague term as it did not specify the difference between this type of highway and other types. Therefore, the Countryside Act 1968 required all highway authorities to reclassify RUPPs as public footpaths, public bridleways or Byways Open to All Traffic. However, not all RUPPs could be reclassified into one of these categories as the reclassification process is very complex and involves checking historical records of how the given route was used as well as enquiries with local people. Therefore, the Countryside and Rights of Way Act 2000 introduced a category of a restricted byway and requested all remaining RUPPs to be reclassified as restricted byways. Although this act was introduced over 10 years ago, there are still some RUPPs which have not been reclassified (Naturenet, 2011).

The public rights of way are established on private land. The land under the path belongs to the landlord, while the surface of the path is the property of the highway authority. The path becomes a public right of way when the owner "dedicates" it to public use. In fact, very few paths have been formally dedicated. However, the law assumes that if the public uses a path for some period of time with no interference, it means that the owner had intended to dedicate it as a right of way. A right of way once established over a path does not cease. There is a legal maxim "Once a highway, always a highway".

The legal consequence of land being a highway is that the public have a right to pass and repass along the route, which means that there is no limit on the frequency of this highway use (Sydenham, 2001, p. 1). It is possible to take a natural accompaniment which includes a pram or pushchair. Wheelchairs can also be used provided the surface of the path is suitable for them. Dogs are allowed on the highways if they stay under the close control of their owners. Straying from a path, or using it for other purposes than passing and repassing, is considered trespassing against the landowner and is a civil offence. The users of a highway may stop to rest or admire the view, or to consume refreshments, providing they stay on the path and do not cause an obstruction (Ramblers, 2011).

Not all paths are public rights of way. Paths crossing public parks, open spaces and other sites to which the public has access are not necessarily rights of way as they comprise public space. Paths which run across land owned by such organisations as the Forestry Commission and the National Trust are available for public use but may not be rights of way. Another category of track which is not a private right of way, is a green lane. The green lane is a physical description of an unsurfaced track. It is usually situated between hedges, ditches or walls. The term does not indicate whether the track carries any rights at all (Sydenham, 2001, p. 6).

There are also permissive paths, which are paths over which landowners allow access without dedicating a right of way. They are often indistinguishable from highways to the users, but there are some important differences:

- A permissive path must have some sign or similar indication that it is not intended to be a right of way.
- The landowner can close off or divert the path if they wish to do so, without any legal process being involved.
- The landowner can make restrictions which would not normally apply to highways, for example to allow horse riding but not cycling, or the other way around.

Permissive paths are commonly found on land owned by a body which allows public access, such as a local authority, a Railway Authority, or the National Trust (Naturenet, 2011).

Public rights of way are signposted and waymarked in the field. Highway authorities have a legal duty to erect and maintain a signpost at every point where a highway leaves a tarmac road. A signpost must indicate the highway status. It may also say where it leads to and give distances to destination points (Ramblers, 2011). The waymarks indicate a line or direction of a path. They are placed on gates, stiles and posts and are based on a colour system: yellow for footpath, blue for bridleway, red for byway and purple for restricted byway. The signposting system is illustrated in Figure 4-19.

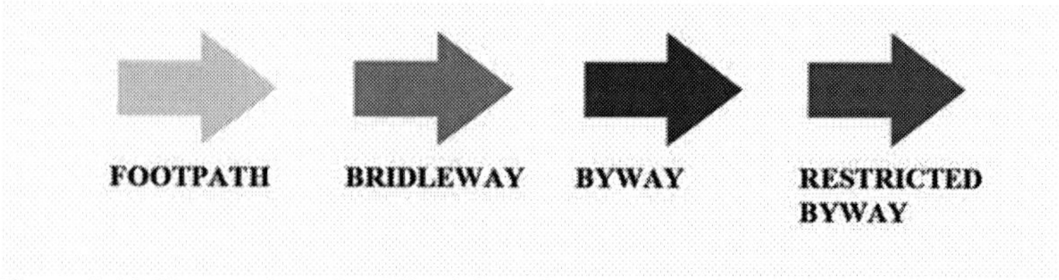

Figure 4-19 Signposting of public rights of way. Image by Ewelina Kwiatek.

The public rights of way within the area are collected in a document called a *Definitive Map and Statement*. It provides the legal record of the public rights to walk, ride or drive on public rights of way and contains particulars as to the position and width of the ways and as to any limitations or conditions affecting the way. It consists of a map and a statement. The map is based on the Ordnance Survey map which may be drawn to a scale of 1:10000. It shows symbols of footpaths, bridleways, restricted byways and byways open to all traffic. The statement which accompanies the Definitive Map may contain particulars relating to the position and width of the path or any limitations or conditions affecting the right of way (Sydenham, 2001, p. 81). These maps may be consulted in order to ensure that a given route is a public right of way. They are available for inspection at the local surveying authorities' offices, in libraries and online. An example of such a map is presented in Figure 4-20, which shows the public rights of way for the Parish of Eythorne in Kent. The footpaths are marked with a purple dotted line, bridleways with yellow lines and byways with red lines. The system of marking rights of way on the Definitive Map is introduced by local surveying authorities and may differ between counties.

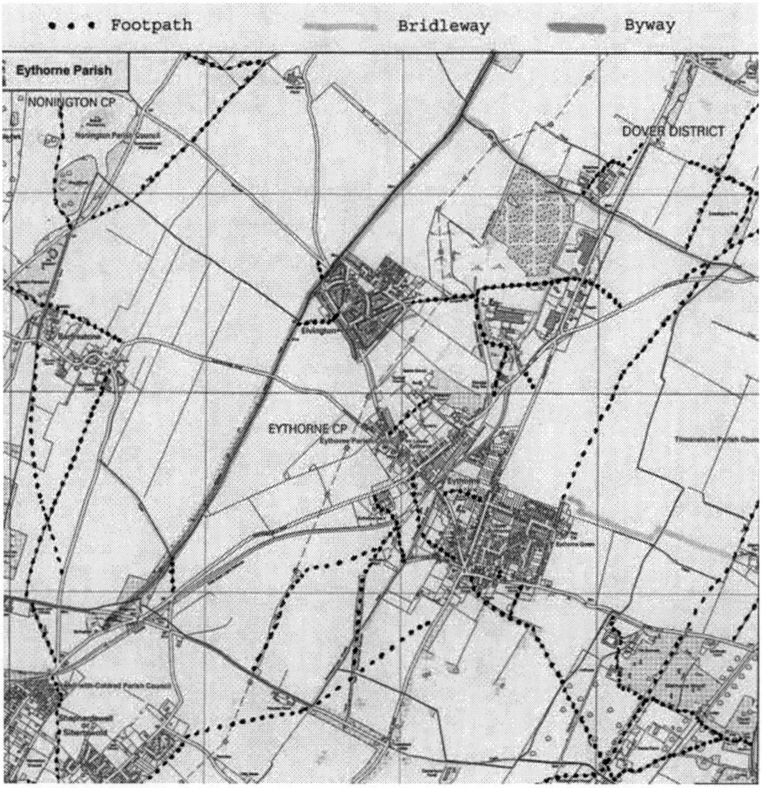

Figure 4-20 Definitive Map of the Parish of Eythorne, Kent. (Kent County Council, 2011)

Polish legislation on rights of way is not as well developed as its British counterpart. Its general approach is different. While in Britain the public right of way relies on the land law, in the Polish legislative system the right of way over the land is mainly concerned with environmental protection. Thus, Polish paths and trails are designated to avoid trespassing on fragile habitats of plants and animals. Furthermore, in England and Wales, public rights of way are established over private land, while in Poland, they are only determined over grounds that belong to the state. They may be designated in national parks and reserves, in city parks, in the countryside or over the part or whole width of the road.

There are very few legal acts in Poland which regulate the public right of way over paths and trails. Legislation on tourist trails is still under development. The most relevant document on the subject is *Ustawa z dnia 16 kwietnia 2004 o ochronie przyrody* 'Legislation act from 16 April 2004 about environmental protection' which states that the public has the right of way on foot, on a bicycle, on skis and on horseback only along trails and ski trails designated by the national park director or the person who is in charge of the national reserve. The act, however, does not provide any definition of *szlak* 'trail' and does not specify types of trails. The web sites of various national organisations, such as *Polskie Towarzystwo Turystyczno-Krajoznawcze* (Polish Association of Tourism and Countryside) and *Lasy Państwowe* (State Forests) list the following types of trails and paths over which the public have the right of way:

- *szlaki piesze* 'foot trails';
- *szlaki i ścieżki rowerowe* 'bike trails and paths', i.e. cycle trails and paths;
- *szlaki do jazdy konnej* 'horse trails';
- *piesze ścieżki* 'footpaths'.

There are no standardised definitions of these types of tourist trails available in any legal document, but there is a general understanding of what these concepts mean in Poland. The difference between *szlak pieszy* 'foot trail' and *ścieżka piesza* 'footpath' is their width and marking. The trail is wider than the path and better marked. There are trails of different levels of difficulty, which are marked with different colours from yellow (the simplest one) to black (the most difficult one) as shown in Figure 4-21 (the most difficult trail represented here is marked with red as there is no black level trail in this area). The only right of way the public have over foot trails and footpaths is on foot.

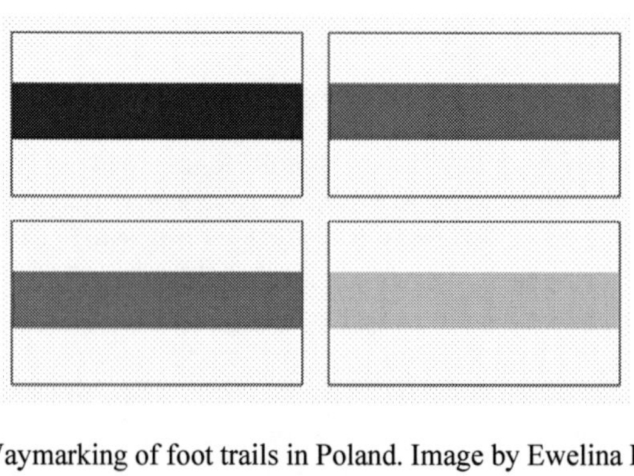

Figure 4-21 Waymarking of foot trails in Poland. Image by Ewelina Kwiatek.

The public have right of way on horseback or leading a horse on horse trails, in addition to right of way on foot. The sign including an orange dot and a horse symbol and showing destination and distance in hours is used to indicate horse trails.

Szlaki and *ścieżki rowerowe* 'cycle trails and paths' are those over which the public have been granted right of way on bicycle, but usually one may also walk along these tracks. They are signposted with the symbol of a bicycle and an arrow showing a direction. The colour of the arrow indicates the difficulty of a trail or path and the colour system is the same as for footpaths and foot trails. The signposts also include information about destination and distance and a cycle trail/path name or number.

The Polish Tourism Organisation (Polska Organizacja Turystyczna, 2011) added *turystyczny szlak samochodowy* 'tourist car trail' to the group of existing trails. A tourist car trail should lead to a tourist attraction and should not run along a motorway or express route. All other types of roads with hard surfaces can be classified as *szlaki samochodowe* 'car trails'. Car trails are signposted with white-brown signs as shown in Figure 4-22.

Figure 4-22 Signposting of car trails in Poland. Photograph by Ewelina Kwiatek.

The concept of a *Definitive Map and Statement* does not exist in Polish, but there are different sorts of maps of tourist trails. An example of such a map with the colour scheme for trails is presented in Figure 4-23.

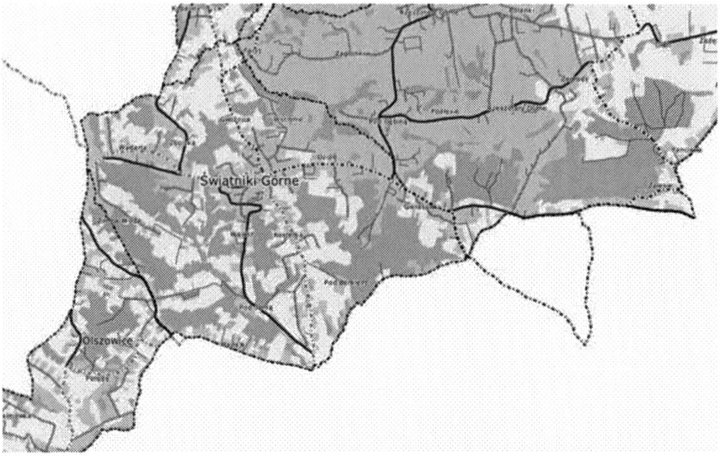

Figure 4-23 The tourist trails in the Świątniki Górne region. Reproduced with permission from Gison Sp. z o.o. (2012).

The map in Figure 4-23 presents tourist trails in the Świątniki Górne region, close to Kraków in the Malopolska province. The green and red dotted lines are used to mark the trails in this region.

The discussion of the concept systems of public rights of way in the UK and Poland highlights various differences which give rise to conceptual mismatches. Although there is a concept of a national trail in English, it is not practical to interpret the English concept system in terms of trails. The *national trail* is defined as a long distance route for walking, cycling and horse riding through the finest landscapes in England and Wales (National Trails, 2011). It is also stated that national trails have been created by linking existing local footpaths, bridleways and minor roads and by developing new ones where there were gaps. The definition of *national trail* is based on public rights of way so interpreting the system of rights of way in terms of trails would lead to circularity.

4.3.8 Case study 8: land registration vs cadastre

Systems of registering rights to land differ between countries. In the UK the system of recording and registering rights to land is called *land registration system*, while in other European countries it is referred to as *cadastre*. The divergences between the two systems result from different land laws and are reflected in distinct registration principles. This discrepancy between the systems engenders conceptual mismatches, which relate to various aspects of land registration. Table 4-17 presents the classification of land registration systems in the Anglo-Saxon countries.

Table 4-17 Land registration system in the UK

term	definition	hyperonym
land registration system	system by which the ownership of estates in land is recorded and registered in order to provide evidence of title and to facilitate dealing Duhaime Organisation (n.d)	
deed registration system	land registration system in which the deed, being a document which describes an isolated transaction, is registered Henssen (1996)	land registration system
title registration system	land registration system in which the legal consequence of the transaction, i.e. the right or the title is registered Henssen (1996)	land registration system
private conveyance system	form of deed registration system used in developing countries, in which most of the transactions cannot be registered and are either formally or informally conveyed Farvacque and McAuslan (1992, p. 56)	deed registration system
Torrens registration system	land registration system invented by Robert Torrens and in which the government is the keeper of the master record of all land and their owners Duhaime Organisation (n.d)	title registration system

The concept of *land registration system* is quite broad as it covers various forms of land registration. The two most common systems of land registration are the deed and the title registration system. In the deed system, the deed itself is registered. The deed is a document confirming that the transaction took place but it is not a proof of the legal rights of parties involved in this transaction. The deed registration system is applied in countries where land registration is mainly based on Roman law, which include France, Spain, Italy, Belgium and the Netherlands. These countries influenced countries in Southern America, parts of North America, Africa and Asia where this system was also adopted (Ghilani & Wolf, 2008, p. 861). This system was never widely adopted in the UK and was eventually dropped. It was replaced with title registration system in the 1860s.

The title registration system focuses on registration of the legal consequence of the transaction that took place, which is the right or the title (Henssen, 1996). Thus, the system registers the right, the name of the rightful claimant and the object of that right with its restrictions and charges.

There are different types of title registration system around the world, which have the same principles but differ in procedure. Henssen (1996) distinguished three groups which reflect different kinds of title registration system:

- the English group, which encompasses such countries as England, Wales, Ireland, some Canadian provinces, Nigeria;
- the German/Swiss group which includes Germany, Austria, part of France (the Alsace-Lorraine province), Switzerland, Egypt, Turkey, Sweden, Denmark;
- the Torrens group with such countries as Australia, New Zealand, some provinces of Canada, some parts of the USA, Morocco, Tunisia, Syria.

These three groups differ not only in land law, but also in mapping/surveying aspects. The English group uses large scale Ordnance Survey maps, the German group employs parcel-based cadastral maps and the Torrens group makes use of incidental survey plans.

The *Torrens Registration system* is listed as a separate concept. The Torrens system is an exampleof a title registration system. In this system, a land title certificate is a sufficient document to show full, valid and indefeasible title (Duhaime Organisation, n.d). The system was introduced by Sir Robert Richard Torrens in 1858 in South Australia, but later it spread to other parts of the world. The Torrens system seems to be influenced by the German land registration in Hamburg and by the German shipping registration (Henssen, 1996).

There is also a concept of *private conveyance system*. This system is common in developing countries. Its aim is to register deeds. However, in such countries as Ghana, Pakistan or Bangladesh only around 10 or 20 percent of transactions are registered. The remaining transactions are either formally or informally conveyed, with or without a person who has legal training. In the past, when communities were very small and people knew their neighbours, oral declaration and symbolic payment were sufficient evidence of the land transfer. Nowadays, conveyancing and registration rules are better developed in those countries and there is a requirement that land transactions must be written and a witness is needed to confirm that a transaction has taken place between the parties (Farvacque & McAuslan, 1992, p. 56).

In the UK the title registration system became the major land registration system. It was gradually expanded and nowadays 90 per cent of titles are registered. It is often referred to as the *land registration system* in the UK (Smith, 2010, p. 110) as it is the one currently applied. Similarly, the formulation *land registration* in Ghana will indicate the private conveyancing system. Thus, *land registration* tends to be used to refer to whichever system is in current usage, even if systems are different.

Smith (2010, p. 114) points out that *land registration* is a misleading term because it is not the land itself as a geographical area that is registered, but a freehold or leasehold estate. Thus, there can be two or more registered titles in relation to the same plot of land: a freehold title and a leasehold title.

The title registration system requires that ownership should be registered, with adverse rights such as estate contracts. The entry must be introduced in the official document called a land register, confirming that the land has the status of the registered land. A register of title is an 'authoritative record, kept in a public office, of the rights to clearly defined units of land as vested for the time being in some particular person or body, and of the limitations, if any, to which these rights are subject' (Simpson, 1976). The register includes such information as the title registration number, definition of the parcel of land, name and address of the owner, and any particulars affecting the parcel that belongs to someone else. Extracts from various parts of title register are available on the Land Registry website.

The register consists of three sections:

- property register;
- proprietorship register;
- charges register.

The property register identifies the geographical location and extent of the registered property by means of a short description (usually the address) and by reference to an official plan which is prepared for each title. It may also provide information on any rights that benefit the land. In the case of a leasehold, it gives brief details of the lease (Land Registry, 2008c). An example from the property register is presented in Figure 4-24.

A: Property register

The register describes the registered estate comprised in the title.

CORNSHIRE : MARADON

1. (19.12.1989) The Freehold land shown edged with red on the plan of the above title filed at Land Registry and being 13 Augustine Way, Kerwick (PL14 3JP).

2. (19.12.1989) The land has the benefit of a right of way on foot only over the passageway at the rear leading into Monk's Mead.

3. (03.12.2003) The exact line of the boundary of the land in this title (between the points A - B in blue on the title plan) is determined under section 60 of the Land Registration Act 2002 as shown on the plan lodged with the application to determine the boundary dated 3 December 2003.

   Note: Plan filed.

Figure 4-24 Specimen from the property register. Reproduced with kind permission of Land Registry. Crown copyright Land Registry (2012a)

The proprietorship register specifies the quality of title and provides such details as the name and address of the legal owner(s) and points out if there are any restrictions on their power to sell or mortgage the property. It may also provide information on the price paid for the title or on the value of the property (Land Registry, 2008b). Figure 4-25 shows an example from the proprietorship register.

B: Proprietorship register

This register specifies the class of title and identifies the owner. It contains any entries that affect the right of disposal.

Title absolute

1. (10.07.2000) PROPRIETOR: PAUL JOHN DAWKINS and ANGELA MARY DAWKINS both of 28 Nelson Way, Kerwick, Maradon, Cornshire PL14 5PQ and of pjdawkins@ail.com.

2. (10.07.2000) The price stated to have been paid on 2 June 2000 was £78,000.

3. (10.07.2000) RESTRICTION: No disposition by a sole proprietor of the registered estate (except a trust corporation) under which capital money arises is to be registered unless authorised by the Court.

4. (05.10.2002) Caution in favour of Mary Gertrude Shelley of 18 Cambourne Street, Kerwick, Maradon, Cornshire PL14 7AR and of Messrs Swan & Co of 25 Trevisick Street, Kerwick, Maradon, Cornshire PL14 6RE.

5. (28.11.2003) RESTRICTION: No disposition of the registered estate by the proprietor of the registered estate is to be registered without a written consent signed by the proprietor for the time being of the Charge dated 12 November 2003 in favour of Fast and Furious Building Society referred to in the Charges register.

Figure 4-25 Specimen from proprietorship. Reproduced with kind permission of Land Registry. Crown copyright Land Registry (2012a)

The charges register contains details of registered mortgages and information on other financial burdens secured on the property (without providing the amounts of money involved). It also gives information on whether the property is subject to any other rights and interests such as leases, or rights of way of covenants restricting the property use (Figure 4-26).

C: Charges register

This register contains any charges and other matters that affect the registered estate.

1. (19.12.1989) The passageway at the side is included in the title is subject to rights of way on foot only.

2. (10.07.2000) A Transfer of the land in this title dated 2 June 2000 made between (1) John Charles Brown and (2) Paul John Dawkins and Angela Mary Dawkins contains restrictive covenants.

   NOTE: Original filed.

3. (01.08.2002) REGISTERED CHARGE dated 15 July 2002 to secure the moneys including the further advances therein mentioned.

4. (01.08.2002) Proprietor: WEYFORD BUILDING SOCIETY of Society House, The Avenue, Weyford, Cornshire CN12 4BD.

5. (28.11.2003) REGISTERED CHARGE dated 12 November 2003.

6. (28.11.2003) Proprietor: FAST AND FURIOUS BUILDING SOCIETY of Fast Plaza, The Quadrangle, Weyford, Cornshire CN14 3NW.

7. (03.12.2003) The parts of the land affected thereby are subject to the leases set out in the schedule of leases hereto.

Figure 4-26 Specimen from charges register. Reproduced with kind permission of Land Registry. Crown copyright Land Registry (2012a)

The Land Registry, when registering a title, assigns a unique reference number to it under which the property details are stored in the Land Registry computer system. This reference number is used in the two types of documents prepared by Land Registry when the title is registered: a *title register*, which has already been discussed, and a *title plan*. The plan shows the land included in the title, which is usually edged in red. It also gives information on the scale at which the plan is drawn. Title plans are prepared on the most recent Ordnance Survey map available at the time of registration and are not updated (Land Registry, 2008a). The title plan for the title number CS72510 (the same one for which the title register was prepared) is presented in Figure 4-27.

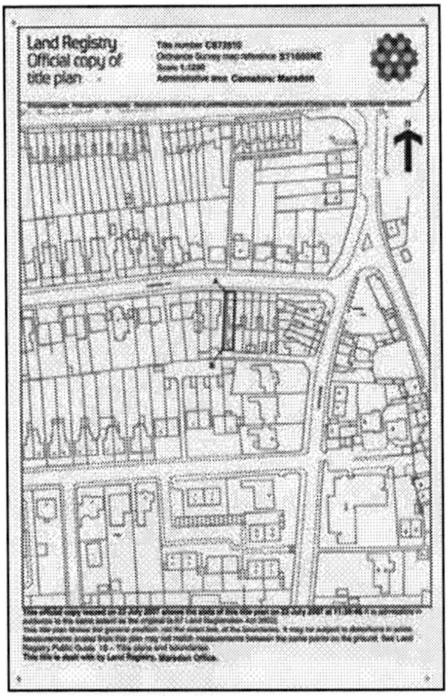

Figure 4-27 Title plan. Reproduced with kind permission of Land Registry. Crown copyright Land Registry (2012b)

The land edged in red is included in the title CS72510. There has been a conveyance of the land tinted pink on the title plan. The colour tinting has been used to show burdens on the title, which include an exception and reservation in favour of the Vendor, who is granted the right to enter upon a land conveyed in order to construct a public sewer. The approximate line of this sewer is shown by a blue broken line on the title plan.

The concept system of registering rights to land in Poland is based on the concept of cadastre. The Polish concept of *cadastre*, although interpreted as *ewidencja gruntów i budynków* 'land and building registry' is in fact a wider concept and it builds on the division of the country into cadastral units. The concept system of the Polish cadastre is presented in Table 4-18.

Table 4-18 System of registering rights to land in Poland

term	definition	hyperonym
kataster 'cadastre'	methodically arranged public inventory of data concerning properties within a certain county or district, based on a survey of their boundaries Gaździcki (2005)	
ewidencja gruntów i budynków 'land and building register'	uniform, national and systematically updated collection of information pieces concerning lands, buildings and premises, as well as their owners and other private or legal persons, who hold those lands, buildings and premises (Kancelaria Sejmu RP, 1989)	cadastre
ewidencja podatkowa nieruchomości 'the register of tax on properties'	registry that is kept by taxation organisations in order to estimate and collect real estate taxes, forest taxes and agriculture taxes Albin (2003)	cadastre
księgi wieczyste 'land and mortgage register'	official registry of material law that is related to real estate; in Poland it is kept by magistrates Gaździcki (2005)	cadastre

The concept of cadastre in Polish is quite complex and, in order to explain it properly, I have to look at the history of cadastre in Poland. The Polish cadastral system dates back to the 18th century. Poland was partitioned and colonised by Germans, Austrians and Russians between 1772 and 1918 and cadastre was kept in German and Austrian partition (Harvey, 2005, p. 293). Therefore, even today, Polish cadastre is still loosely based on German cadastre. Unlike the German one, however, it is separated into three unrelated registers: *ewidencja gruntów i budynków* ('register of land and buildings', i.e. land and building registry), *księgi wieczyste* ('eternal books', i.e. land and mortgage register) and *ewidencja podatkowa nieruchomości* ('register of tax on real estates', i.e. register of tax on properties). The three registers are kept by three different ministries and at the moment they are not coordinated, although various studies have been undertaken to integrate them, e.g. Albin (2003). The land and building registry is under the control of the office of the Surveyor General of the Polish Ministry for Infrastructure. This registry documents all land surveying activities as the relevant documents are sent to the surveying offices first at local government level, and then on to the district level. The land and mortgage register is under the control of the Justice Ministry and requires all property transactions to be recorded in the district courts where the registers are kept. This register is not very efficient. Due to the complex taxation system with which Polish courts have to deal when registering properties, delays in registration of up to two years, and even more, occur (Harvey, 2005, p. 298). The real estate tax register is managed by the Ministry of Finance and is used to determine and assess a property tax.

The land and building register is the most efficient of the three registers and it is associated with the cadastre in Poland (Ustawa z dnia 17 maja 1989 r. Prawo geodezyjne i kartograficzne). This designation, however, seems to describe what the cadastre should be, rather than what it is actually is in Poland. Therefore, referring to the land and building register as cadastre in Poland is actually to oversimplify the facts.

The Polish cadastral system relies on the division of the country into cadastral parcels, cadastral precincts and cadastral units, which were created specifically for the purpose of running an efficient cadastral system in Poland. They are described in Table 4-19.

Table 4-19 Cadastral division of Poland

term	definition	holonym
jednostka ewidencyjna 'cadastral unit'	commune or place separated from the commune as a town/city or city district that function in the Polish registry of lands and buildings as units of territorial division of the country Gaździcki (2005)	
obręb ewidencyjny 'cadastral precinct'	unit of a surface division of a country that was created for the purpose of land and buildings recording; it has its own recording frame, consists of recorded plots and coincides with the boundaries of villages and *sołectwo* (an administrative unit in Poland, may comprise of part of a village, one or more villages) Okła (2009)	jednostka ewiden-cyjna
działka ewidencyjna 'cadastral parcel'	continuous land area that is situated within one precinct, homogeneous from the legal point of view and separated from the surrounding area by means of borderlines (RMRRB, 2001)	obręb ewiden-cyjny

The cadastral parcel is the lowest level in this division. It is a part of a cadastral precinct, which, in turn, is a part of a cadastral unit. Cadastral units were created because the technique for setting up and keeping land and building registers required them to emerge. A cadastral precinct is the area of land created by cadastral division of the country for which the documents, called *operaty ewidencyjne* 'cadastral statements' are kept (RMRRB, 2001). The land and building register in Poland contains information about:

- grounds, their location, borders, area, type of use, soil class, marking of land and mortgage registers or collections of documents, if they are kept for parcels that include these grounds;
- buildings, their location, designation, utilisation and general technical data;
- premises, their location, utilisation and area.

The land and building register also includes information about the owners of private grounds and persons who are in charge of state and corporation grounds, buildings and building parts and the addresses of these people. The register should also specify the property value and should contain information on whether the buildings are on the register of historic buildings.

Farm and forestry grounds are classified on the basis of the soil-based classification system. The land and building register as well as the soil-based classification system are managed at the local government level by *starosta* 'the office of the prefect of the district'. Courts and notary offices send legal copies of decisions and statements in addition to copies of notary acts, which indicate changes that have been made in the register, to the office of the prefect no later than 30 days from the date when the relevant decision or statement was made or the notary act was issued.

All the information covered in the land and building register are included in *operat ewidencyjny* ('statement of cadastre', i.e. the cadastral statement). The statement is kept for each cadastral precinct (RMRRB, 2001). It consists of an electronic database as well as documents which justify entries made in the database. The database facilitates visualisation of data in the form of registers, files and lists as well as cadastral maps, and provides an opportunity for interested parties to access extracts from these documents. Everyone can access the cadastral statement and request *wypis z rejestru gruntów* ('an extract from the land and building register') or *wyrys z mapy ewidencyjnej* ('an extract from a cadastral map'). An example of such a document is shown in Figure 4-28.

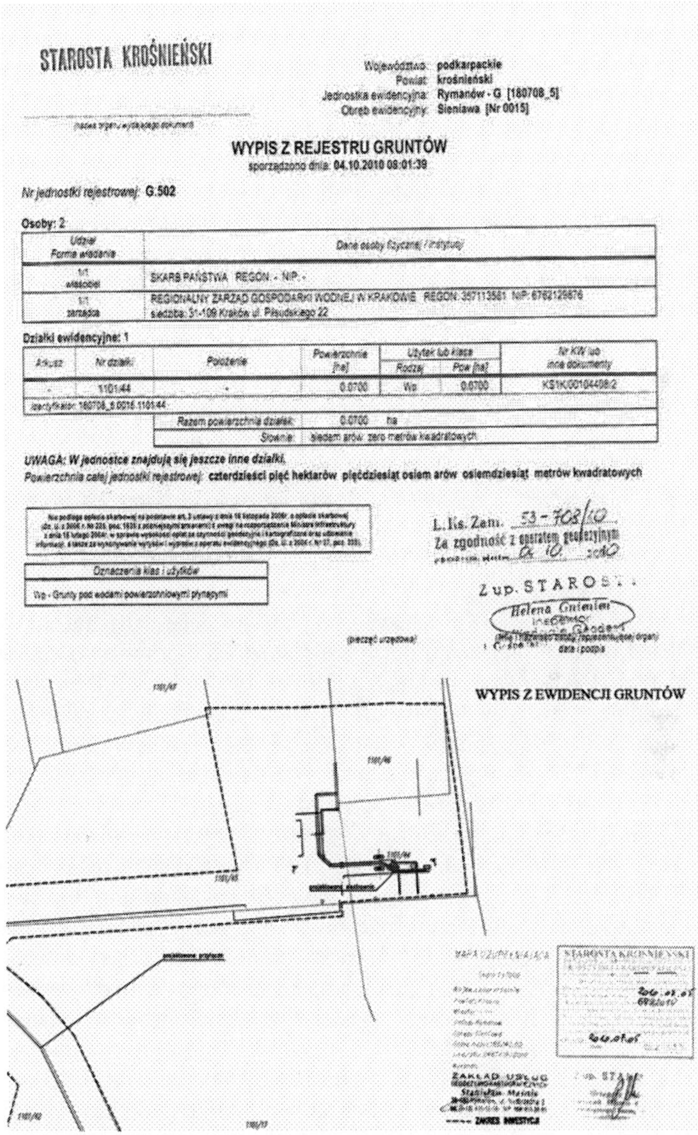

STAROSTA KROŚNIEŃSKI

Województwo: podkarpackie
Powiat: krośnieński
Jednostka ewidencyjna: Rymanów - G [180708_5]
Obręb ewidencyjny: Sieniawa [Nr 0015]

(nazwa organu wydającego dokument)

WYPIS Z REJESTRU GRUNTÓW
sporządzono dnia: 04.10.2010 08:01:39

Nr jednostki rejestrowej: G.502

Osoby: 2

Udział Forma władania	Dane osoby fizycznej / instytucji
1/1 właściciel	SKARB PAŃSTWA REGON: - NIP: -
1/1 zarządca	REGIONALNY ZARZĄD GOSPODARKI WODNEJ W KRAKOWIE REGON: 357113581 NIP: 6762129876 siedziba: 31-109 Kraków ul. Piłsudskiego 22

Działki ewidencyjne: 1

Arkusz	Nr działki	Położenie	Powierzchnia [ha]	Użytek lub klasa: Rodzaj	Użytek lub klasa: Pow [ha]	Nr KW lub inne dokumenty
-	1101/44	-	0.0700	Wp	0.0700	KS1K/00104408/2
identyfikator: 180708_5.0015.1101/44						
		Razem powierzchnia działek:	0.0700	ha		
		Słownie:	siedem arów zero metrów kwadratowych			

UWAGA: W jednostce znajdują się jeszcze inne działki.
Powierzchnia całej jednostki rejestrowej: czterdzieści pięć hektarów pięćdziesiąt osiem arów osiemdziesiąt metrów kwadratowych

Oznaczenia klas i użytków
Wp - Grunty pod wodami powierzchniowymi płynącymi

L. Ks. Zam. 53 - 708/10
Za zgodność z operatem geodezyjnym
04. 10. 2010

Z up. STAROSTY
Helena Gnielec
Inspektor

(pieczęć urzędowa)
data i podpis

WYPIS Z EWIDENCJI GRUNTÓW

ZAKŁAD USŁUG
ZAKRES INWESTYCJI

Figure 4-28 Extracts from a land and building register and from a cadastral map. Reproduced with permission from Ośrodek Usług Inżynierskich STAAND Sp. z o.o. (2012)

Figure 4-28 shows the parcel identified with the number (*numer działki*) 1101/44, which is situated in the cadastral precinct (*obręb ewidencyjny*) 0015, Sieniawa being a part of the cadastral unit (*jednostka ewidencyjna*) 180708_5, Rymanów - G. The extract from the land register also includes such information as area, kind of use, soil class, data for land tax, such as the name of the owner and his address. Usually, it also includes a reference to the land and mortgage register. The extract from the map shows the outline of the parcel and its location in reference to other parcels.

The Polish cadastre dates back to the 17th century, when Poland had a monarchy. It does not have uniform roots, however, as, due to partitions of the country between 1772 and 1795, the type of cadastre depended on the land policy of the country that controlled a given part of Poland. In the part of Poland under the Russian regime the mortgage law was introduced. In Galicia, which belonged to Austro-Hungary, ground books were kept, while in the Prussian part the land and mortgage register based on three legal acts issued between 1897 and 1899 was in use. After 1918 attempts were made to create Polish cadastre, but, due to the fact that existing registers differed greatly or were missing, those attempts were not successful (Kunach, 1999).

Projects to unify Polish cadastre and to introduce cadastral tax were completed by 1938 but could not be implemented due to the outbreak of World War II. They were resumed as late as 1955 after a bill on the land and building register was passed. Cadastre was finally established in Poland between 1956 and 1970 (Kozłowski, 1997). It is uniform for the whole country now, but it is a complex system consisting of three different parts, which are after-effects of its diversified roots.

The land and mortgage registers, which are not very common in other countries, are particularly interesting. They were kept by notary offices until 1991, but this did not prove very efficient, and for this reason, departments of regional courts relevant to the real estate location have taken over these responsibilities (Harvey, 2005, p. 298). They are kept for real estate, not for their owners, which means that one name may appear as the name of the owner in a few registers, while the ownership should be indicated only in one register.

The land and mortgage register consists of four parts. Part 1 includes the section on the designation of the property, which comprises the data from the land and building registers and the list of rights related to a given real estate. Part 2 includes information on the owner(s) or perpetual leaseholder. Part 3 is about restricted property laws (excluding mortgages), restrictions on managing the property and limitations on perpetual lease. Part 4 covers all the issues concerning mortgages.

The register of tax on properties incorporates information which is essential to calculate and collect tax on real estate, agricultural tax and forestry tax. It includes data about tax payers, specifically their name(s), address and tax reference number. It also contains data on the subjects of taxation. The subject of taxation is a parcel, a building or a part of the building, the register provides their area, the parcel/building identifier, the number of the land and mortgage register or the name of the court which keeps the register that is relevant to the particular parcel or building (Ustawa z dnia 6 lipca 1982 r. o księgach wieczystych i hipotece).

As it may be seen from this discussion, the components of Polish cadastre are linked to one another. The land and building register and land and mortgage register are correlated. The land and building register contains the identifier of the land and mortgage register, which is specific to a given property. The land and mortgage register, on the other hand, includes the designation of the property as an excerpt from the land and building register in its Part 1. The register of tax on properties is related mainly to the land and building register although it also includes *numer księgi wieczystej* 'number of the land and mortgage register' for the particular property.

To sum up this part of the discussion of different systems of registering rights to land in the UK and in Poland, I may see that the British title register shares many features with the Polish land and mortgage register, as they both focus on proprietorship, property laws and charges. The Polish land and building register contains data which is not specified in the British land register, such as the type of land use and the soil type (not to mention the fact that the British and Polish classifications of soil types are different). The Polish cadastral system seems to be very complex when compared to the British system. The register of tax on properties and the land and building register do not seem to have corresponding registers in the British system.

The management of registers is also very different between the two countries. In Poland, the three different registers are managed by three different institutions. The land and building registers are kept by the office of the prefect of the district, the land and mortgage registers are under the control of courts and the register of tax on properties is managed by taxation bodies.

In the UK, land administration institutions have developed in a different way from the rest of Europe. There is no single organisation responsible for the cadastre, but there are various organisations responsible for the recording of land rights, such as Her Majesty's Land Registry (HMLR – England and Wales), Registers of Scotland (RoS) in Scotland and land Registers of Northern Ireland. The organisations in charge of land and property valuation are the Valuation Office Agency (for England and Wales), the Scottish Assessors Association in Scotland, and the Valuation and Land Agency in Northern Ireland (Permanent Committee on Cadastre in the European Union, 2011).

There are also variations with regards to mapping, which is part of the title registration process in the UK, where title plans are created, and a part of cadastral process in Poland, with cadastral maps being generated. Both title plans and cadastral maps are large-scale elaborations. They differ in a few respects, however. The Title Plan shows an outline of the property and its location in relation to local properties and North. It may also contain 'T' markings to show the location of fences. Party walls may also be highlighted. It may contain coloured markings referred to in the Title Register for rights of way and boundary lines (Land Register Online, 2008). Cadastral maps usually include additional information on land use and type of soil.

Another difference between most mainland European countries and the UK is base mapping (Permanent Committee on Cadastre in the European Union, 2011). Base mapping in the UK is topographic, which means that it shows features that exist on the ground but does not show fixed boundary points and monuments. On the contrary, base maps in Poland include information about cadastre, land management (streets, trees, public utility objects, underground, surface and overhead land infrastructure (pipes, cables, etc) and lay of the land (Bucewicz et al., 1981).

Various organisations are in charge of these maps. Ordnance Survey, as a national mapping agency, maintains large scale mapping for England, Scotland and Wales. In Northern Ireland, Ordnance Survey Northern Ireland is responsible for this task. These two government agencies maintain the detailed digital maps which provide a definitive framework upon which various organisations can manage their data (Permanent Committee on Cadastre in the European Union, 2011). In Poland, the base map is maintained by Ośrodki Dokumentacji Geodezyjnej i Kartograficznej 'geodetic and cartographic documentation centres' (Instrukcja techniczna K - 1. Mapa Zasadnicza, 1981).

To conclude this section, I realise that conceptual mismatches are quite frequent in surveying terminology and that they, indeed, result in lexical gaps that need to be solved. In the next section, I will provide a general overview of translation strategies for dealing with conceptual mismatches, and in the final section of this chapter I will try to provide solutions to the case studies discussed in this section by matching translation strategies with particular problems.

4.4 Translational strategies for dealing with conceptual mismatches

Conceptual mismatches occur when the concept systems of two languages differ. As a consequence of this variation, a target language expression differs in meaning from any corresponding expression in the source language. The terms *lexical gap* (Hann, 2004, Lyons, 1977; Janssen, 2004) and *translation mismatch* (Prahl & Pretzolt, 1997, Pause, 1997; Kameyama et al, 1991) are used to describe cases where it is impossible to render the exact source text meaning in the target language. It is quite important to note that the latter designation has been used only in the context of machine translation, whilst the term *lexical gap* is used very widely and has been applied in technical translation (Hann, 2004), in the context of multilingual lexical databases (Janssen, 2004) and in semantics (Lyons, 1977). Translators, in general, do not have a separate designation for this case and call it *non-equivalence* (Newmark, 1988, Baker, 1992). I will first analyse how these three designations differ, and then I will focus on strategies which are used to deal with them.

The notion of a *translation mismatch* refers to situations where there are actual differences in the information that is conveyed from the source to the target language. If the source text is vague or ambiguous in a way that cannot be rendered in the target language, the translation process must add information by making a guess about the intent of the source text and then rendering that intent into the target text. If, on the other hand, the target language allows a degree of ambiguity or vagueness that is not allowed on the level of the source text, it may be necessary to remove some information from the source text and not render the source text completely in the target language. For example, the English word *wood* has two translations in Polish: *las* and *drewno*, depending on whether it is a collection of trees or a substance. Thus, the translation from English into Polish requires additional information, while going from Polish into English involves omitting some information which is present in Polish.

What Kameyama et al. (1991) understand as translational mismatches, Lyons (1977), Janssen (2004) and Hann (2004) interpret as lexical gaps. Lyons (1977, p. 301) states that a lexical gap is the absence of a lexeme at a particular place in the structure of a lexical field. For example, Polish has the word *szwagier* to refer to wife's brother, husband's brother, and sister's husband; the word *szwagierka* for wife's sister, husband's sister;

and *bratowa* for brother's wife. There are no words for *brother's husband* and *sister's wife* in Polish. In English, with the in-law vocabulary, this problem can be easily resolved. However, in Polish these concepts are holes in the lexical structure and constitute lexical gaps.

Janssen (2004, p. 137) understands lexical gaps as words for which there is no direct translation in the target language. For example, the English word *toddler* would be translated into Polish as *dziecko* 'child', but *dziecko* is not a complete translation, since *toddler* means a young child, especially one who is learning or has recently learned to walk (Cambridge Advanced Learners' Dictionary, 2011). There is no single word for *toddler* in Polish and therefore it is justified to say that there is a lexical gap in Polish for this word.

Hann (2004, p. 182) claims that different conceptual structures in the languages concerned result in different terminological structures, which, in turn, lead to conceptual mismatches that are reflected in lexical gaps.

This view on lexical gaps, which interprets them as synonymous with translation mismatches, is inconsistent with the presentation of lexical gaps by Bentivogli and Pianta (2000, p. 664), who delineate lexical gaps as one type of idiosyncrasy that can occur between pairs of languages. They discuss lexical gaps in the context of a MultiWordNet, which is a project aimed at building a multilingual lexical database that has already been discussed in (4.2.2). They interpret lexical gaps as lexicalisation differences that occur when a language expresses in a lexical unit what the other language expresses with a free combination of words, e.g. *overnight* 'przez całą noc', i.e. through the whole night. Other types of lexical divergences specified by Bentivogli and Pianta include:

- syntactic divergences, which occur when the translation equivalent (TE) does not have the same syntactic ordering properties as the source language word, e.g. *take away meal* in English and *posiłek na wynos* 'meal for carrying' in Polish;
- divergences in connotation, which occur when the TE fails to reproduce all the nuances expressed by the source language word, e.g. *black person* in English and *murzyn* 'Negro' in Polish. *Murzyn* has a neutral connotation in Polish and refers to a black person, while in English *Negro* is offensive. On the contrary, *czarnuch* 'blackey' has a negative connotation;
- denotational differences, which occur when the denotation of the source language word only partially overlaps with the denotation of the TE, e.g. *klasztor* in Polish incorporates both *monastery* (for monks) and *convent* (for nuns) in English.

If this classification of lexical idiosyncrasies is compared with what Janssen (2004) and other linguists claim on the subject of lexical gaps, it may be noticed that, to a great extent, these classifications overlap. However, the interpretation of lexical gaps proposed by Bentivogli and Pianta (2000) seems to be more explicit as it lists different types of lexical divergencies.

Conceptual mismatches thus occur when there are no translation equivalents (TEs) between the source and target languages. They are usually discovered by translators who try to find equivalents for the source language words they translate into the target language by looking at the meaning of the source language words and the meaning of possible translation equivalents. As I am dealing with conceptual gaps in terminology, I will look at non-equivalence at the level of naming unit. An overview of the most common problems of non-equivalence at word level may be found in Baker (1992). In my discussion of this classification, I will focus only on those types of non-equivalence that are relevant to terms and termbases. Thus, I will not consider the expressive dimension of equivalence and non-equivalence that applies to the use of words in the sentence. Non-equivalence most often occurs in the following situations:

- A translator is dealing with culture-specific concepts, which incorporate concepts that are totally unknown in the target culture such as religious beliefs, social customs or types of food. An example of such a concept is *airing cupboard* in English, which is unknown to speakers of many other languages.
- The source language concept is not lexicalised in the target language. This occurs when a concept in the source language is known in the target language but is not lexicalised, e.g. *submission* understood as the 'act of giving a document for a decision to be made by others', has no ready equivalent in Polish.
- The source language word is semantically complex, which means that a single word consisting of a single morpheme expresses a more complex set of meanings than a whole sentence. An example of such a complex word in Polish is *sękacz*, which means a 'popular Polish-Lithuanian sponge-shortbread cake, baked by painting layers of dough onto a rotating spit in a special open oven or over an open fire'.
- The source and target languages make different distinctions in meaning. The number of distinctions made in the source language and the target language may differ, e.g. English makes a distinction

between *finger* and *toe*, while Polish has only one word for both, which is *palec*.

- The target language lacks a superordinate, although it may have all the hyponyms of the given word. For example, there is no equivalent in Polish for the English superordinate *facilities*, meaning 'the buildings, equipment and services provided for a particular purpose'. Nevertheless, all the hyponyms are present in Polish, *budynki* 'buildings', *usługi* 'services', *wyposażenie* 'equipment'.
- The target language lacks a specific word (hyponym). For example, under *jump,* English has more specific words such as *leap*, *vault*, *spring*, *bounce*, *dive*, *plunge* and *plummet*. Polish does not make such a distinction regarding types of jumping and has only the superordinate *skakać* 'jump'.
- There is no equivalent in the target language for a particular form in the source text. This often happens in the case of affixation processes. Some English prefixes and suffixes have no direct equivalents in some other languages, e.g. *employee*, *trainee*, *payee*.

The first sign of a conceptual mismatch is usually the impossibility of finding an equivalent for the source language term in the target language. Analysis of the concept systems both in the source and in target languages confirms whether the conceptual mismatch actually occurs and on what level in the concept structure it may be found.

Parallel tree diagrams (Figure 4-29) are used widely in translation and illustrate lexical gaps well.

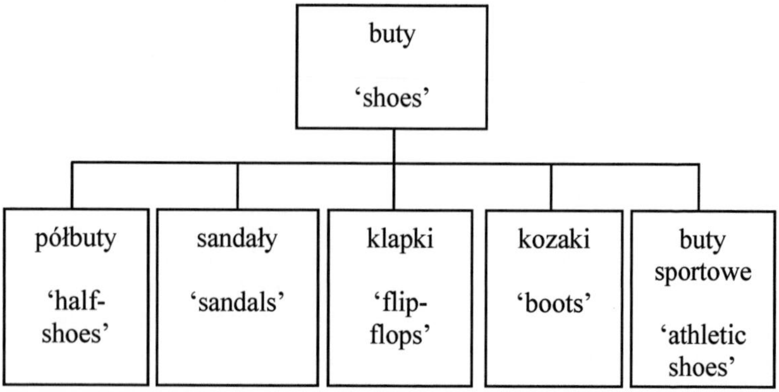

Figure 4-29 Parallel tree diagram. Diagram by Ewelina Kwiatek.

From this diagram, it may be recognised that there is a problem with the word *buty* 'shoes' in Polish, which is a general category including different types of footwear. It includes *kozaki* 'boots', which in English are a separate category from shoes. Therefore, the word *buty* should be translated as *footwear* rather than *shoes*. Another problem concerns the word *półbuty*, which does not seem to have an equivalent in English, and the diagram indicates that the word should be translated by its hyperonyms, thus as *shoes*. In fact, these are synonymous words and if the English equivalent for *buty* is replaced with *footwear* and the English equivalent os provided in the form of *shoes* for *półbuty* the whole concept structure will be correct.

The type of strategy used to deal with conceptual mismatches differs depending on the area in which it is used. Machine translation, multilingual lexical databases, linguistics and translation have their own ways of analysing and solving conceptual mismatches. The common feature of many areas is the use of componential analysis with various modifications to analyse the concept system to which a problematic concept belongs.

In machine translation to analyse the concept system Kameyama et al. (1991) propose using componential analysis with a *distributive lattice of infons*, which is a lattice in which features of the related concepts are described by means of binary elements. Infons in this lattice are elements that uniformly represent information from different levels of linguistic abstraction involving morphology, syntax, semantics, and pragmatics. Thus, a lattice is comprised of linguistic knowledge.

Words in this lattice are associated with properties, for example, the English words *painting*, *drawing* and *picture* have properties which are called P1, P2 and P3 respectively. Infons in the lattice are denoted as <<P, x1, …,xn; 1>> if they stand in the relation P, and as <<P, x1, …, xn; 0>> if they do not stand in the relation P. A string in the "Picture" sublattice for different forms of presentation (i.e. pictures, drawings, paintings, etc.) is linked to a property with the SIGNIFIES relation (indicated as = =) and property as such is interlinked with the INVOLVES relation (indicated as =>) and is represented as follows:

EN: "picture"= = P1 <<picture, x; 1>>
EN: "painting"= = P2 <<painting, x; 1>>
EN: "drawing"= = P3 <<drawing, x; 1>>
EN: "oil-painting"= = P4 <<oil-painting, x; 1>>
EN: "water-colour"= = P5<<water-colour, x; 1>>
P2 => P1, P3 => P1, P4 => P2, P5 => P2

The lattice is similar to a semantic network (Kameyama, Ochitani & Peters, 1991, p. 195). However, it may also play the role of a general translational framework as it facilitates enriching the lattice created for one language with the knowledge of other languages. New infons can be inserted in appropriate places and more instances of the "signifies" relation can be added to the lattice.

The lattice may be extended to cover Polish words for *picture*. In Polish *obraz* includes different types of paintings, as it indicates a piece of art that has been painted. It does not include drawings and photographs, however. Thus, *obraz* seems to be a more specific term than *picture* in English. The Polish word *rysunek* is more specific than the English word *drawing* as it indicates a sketch, a work that is inferior to *obraz* and is not a hyponym of *obraz* in Polish. English words *oil-painting* and *water colour* have equivalent terms in Polish which are *obraz olejny* and *akwarela*, respectively.

The lattice for picture may be extended with Polish words as follows:

EN: "painting" PL: "obraz" = = P6 <<obraz, x; 1>>
EN: "drawing" PL: "rysunek" = = P7 <<rysunek, x; 1>>
PL: "(obraz olejny)" = = P4 <<oil-painting, x; 1>>
PL: "(akwarela)" = = P5<<water-colour, x; 1>>
P5=> P2, P4 => P2 , P6 => P2, P7 => P3

The lattice of infons has a similar role to componential analysis in helping to identify conceptual mismatches as well as in finding a translation with the desired property. Infon lattices are ordered by involvement relations including known real-world constraints. If a given infon is a fact in some situation, all infons higher than that infon in the lattice must also be facts in the situation. Therefore, an accurate translation may be found by looking at the lowest infons.

Kameyama et al. (1993, p. 92) claim that, because languages differ in concepts and real-world entities, translation often involves an approximation of the source text meaning rather than finding an exact counterpart equivalent in the target language. When the target language lacks a word or a phrase for something in the source text, translation must either generalise the concept mentioned or specialise it. Generalisation involves omitting some information the source text contains, while specialisation requires recovering implicit information from the source language context. Generalisation is regarded as a safer translation strategy because it is supposed to preserve accuracy of information. If it is used, it should be slight as otherwise a good deal of information will be lost (Kameyama, Peters & Schütze, 1993, p. 95).

Prahl and Petzolt (1997, p. 137), who also work in the field of machine translation, consider conceptual mismatches as translation problems occurring when there is an information deficit in a certain context at a specific moment within the translation process. They offer similar solutions to translation mismatches:

- reduction strategies, which are used when the missing information is not relevant at a given point in time, and the translator can use a technique such as generalisation;
- achievement strategies, which are applied when the missing information is relevant at a given point in time, and the translator tries to obtain more information.

Their strategies correspond to generalisation and specialisation strategies by Kameyama et al. (1993).

Janssen (2004) offers a slightly different approach as he is concerned with the problem of conceptual mismatches between languages in the design of Multilingual Lexical Databases (MLLDs). His work is particularly relevant to my research as it is the only publication I found on lexical gaps in the context of lexical databases. He works with English, Spanish and French. The example of *dedo* in Spanish vs. *finger* and *toe* in English, which constitute the conceptual mismatch, may be well applied for Polish, where there is a single word *palec*, meaning any of the five separate parts at the end of the hand and foot, which refers both to *finger* and *toe* in English.

Janssen (2004, p. 138) provides an overview of existing approaches used to deal with lexical gaps such as the one mentioned above. In his view, there are lexical gaps both in English as there is no corresponding word for *palec*, and in Polish as there are no direct equivalents for such specific words as *finger* (*palec u ręki* 'finger/toe at hand') and *toe* (*palec u nogi* 'finger/toe at foot').

These lexical gaps may be solved using the project-down approach and the hyperonymic approach. In the project-down approach, the sense of the hyperonym *palec* is discarded and replaced by the two more specific meanings of *finger* and *toe* as shown in Figure 4-30. The distinction between the sense of *finger* and the sense of *toe* is hence introduced into Polish, removing the lexical gap (the use of capital letters indicates concepts). Thus, the lexical gap in Polish is removed by duplication.

$$\text{palec}_1 \rightarrow \text{FINGER} \leftarrow \text{finger}$$
$$\text{palec}_2 \rightarrow \text{TOE} \leftarrow \text{toe}$$

Figure 4-30 The project-down approach after Janssen (2004, p. 138)

Although the project-down approach solves the problem of lexical gaps, it has some deficiencies. It introduces an ambiguity in Polish as it duplicates forms. There is only one form in Polish, which is *palec* which is the label for one concept in Polish. This concept corresponds to two concepts in English: finger and toe. A database using the project-down method is difficult to maintain as each time new languages are added, new distinctions in meaning have to be considered and implemented.

The second method is called the hyperonymic approach and is illustrated in Figure 4-31. In this approach, the word *palec* is modelled as a hyperonym of the words *finger* and *toe*. This solution simply acknowledges the existence of the lexical gap. It neither fills the gap nor removes it. But they are resolved at a later stage.

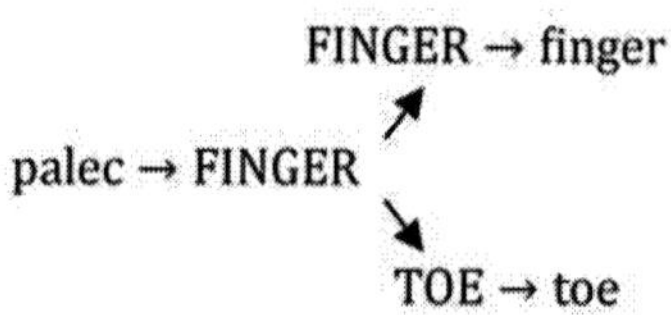

Figure 4-31 Hyperonymic approach after Janssen (2004, p. 139)

The hyperonym approach has three variants: the variant with no interlingua, the variant which uses an unstructured interlingua and the third variant that uses a structured interlingua (Janssen, 2004, p. 141). By interlingua, Janssen understands a concept system that will be common for all the languages involved in the process of the identification of conceptual mismatches.

In the variant with no interlingua, the hyperonymy links are present between the language-dependent word-senses as shown in Figure 4-32.

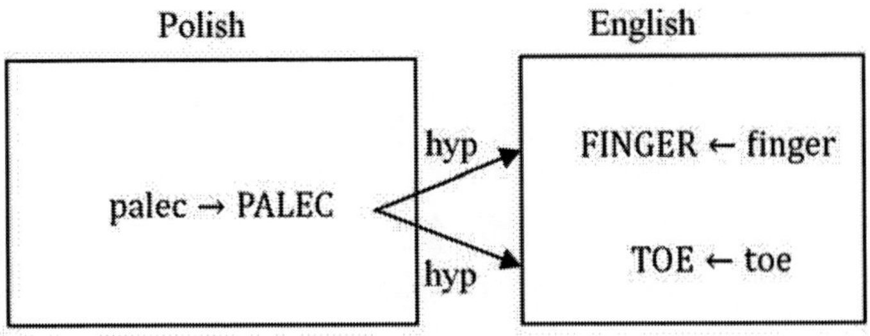

Figure 4-32 No Interlingua after Janssen (2004, p. 141)

The example of the hyperonymic model with no interlingua is the Hub-and-Spoke model (Beeken et al., 1998), in which one language has the role of a hub and all other languages are playing the role of spokes. In this model, every hyperonymy link conveys the appropriate differentiating information. So, in the case of *finger*, the hyp-link is marked by ‘u ręki’ to indicate that a finger is a *palec u ręki*.

In the hyperonymic model with unstructured interlingua shown in Figure 4-33 the hyperonymy links are between the words (or word-senses) and the interlingual concepts. An example of such a system is EuroWordNet (EWN). The interlingua in EWN consists of a list of Interlingual Items (ILIs) to which all the synsets of the WordNets of the various languages are linked.

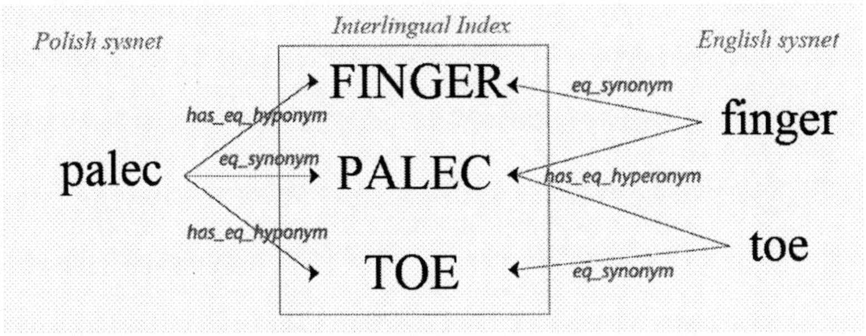

Figure 4-33 EuroWordNet after Janssen (2004, p. 142)

The problem with this interlingua is that the hyperonymy relations are not part of the interlingua, but are between the interlingua and various languages which means that the word *palec* has to be linked to three different ILIs. When there are more languages involved, the number of links for a single word (synset) can increase rapidly. As the hyperonymy relation is between the languages and the interlingua, the hyperonymy needs to be re-established for each individual language.

The system with structured interlingua seems to provide the most convenient solution. In this set-up, the hyperonymy links are between the various meanings in the interlingua (Figure 4-34).

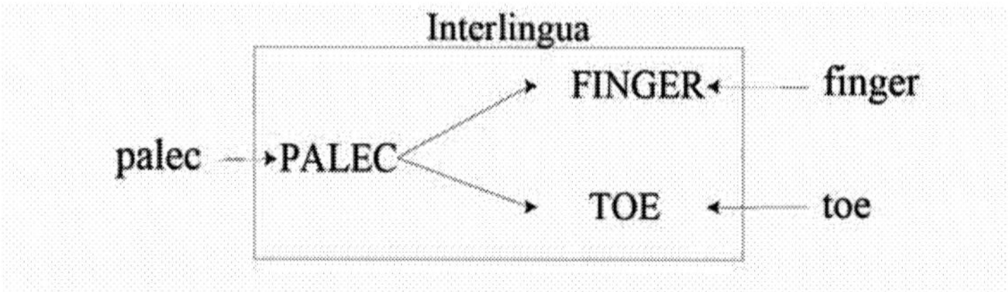

Figure 4-34 No Interlingua after Janssen (2004, p. 143)

An example of a structured interlingua system is a Structured Interlingua MultiLingual Lexical Database Application (SIMuLLDA) proposed by Janssen (2002) in his thesis. In SIMuLLDA, every word of every language relates to as many interlingual meanings as it has senses and the interlingual meanings are related hierarchically. The meaning PALEC is a hyperonym of both FINGER and TOE. FINGER has a definitional attribute labeled as *hand*, whereas TOE has a definitional attribute *foot*. These attributes are parts of the interlingua structure and therefore they are themselves lexicalised.

SIMuLLDA aims to provide a tool for lexicographers to support the generation of bilingual dictionaries. The core of SIMuLLDA consists of dictionary data transformed into a structured hierarchy by means of logical tools (Janssen, 2004, p. 144). The dictionary data for different words to refer to horses including citation forms and definitions are given in Table 4-20. The example was originally provided in English and a comparison was made with French to illustrate conceptual gaps between the two languages. However, I will apply it to Polish which distinguishes similar types of horses according to their sex and age to French.

Table 4-20 Definition of words for horses after Janssen (2004, p. 144)

colt	a young male horse
filly	a young female horse
foal	a young horse
mare	a fully-grown female horse
stallion	a fully-grown male horse

The definitions in Table 4-20 are analysed as items that relate English words to defining aspects of the meaning that is expressed by these words. These defining aspects are called definitional attributes. Thus, the first definition relates *colt* to the definitional attributes *young* and *male*. *Colt* is also related to *horse*. Table 4-21 renders a complete presentation of the dictionary definitions as features that relate English words and definitional attributes.

Table 4-21 Analysis of definitions of horses after Janssen (2004: 145)

	horse	male	female	adult	young
horse	X				
colt	X	X			X
filly	X		X		X
foal	X				X
mare	X		X	X	
stallion	X	X		X	

The rows in Table 4-21 present the interlingual meaning of words as the definitional attributes refer not only to the source language, which is English, but also to Polish, the target language. The data in Table 4-21 is transformed into an interlingual structure, which Jannsen (2004) calls the *concept lattice*. The conventions adopted in this lattice are as follows: word forms are type set in Cambria, definitional attributes in **bold face**, and interlingual meanings (concepts) in SMALL CAPS. The concept lattice is illustrated in Figure 4-35. It was created for English and French but I modified it so that it can be used for Polish.

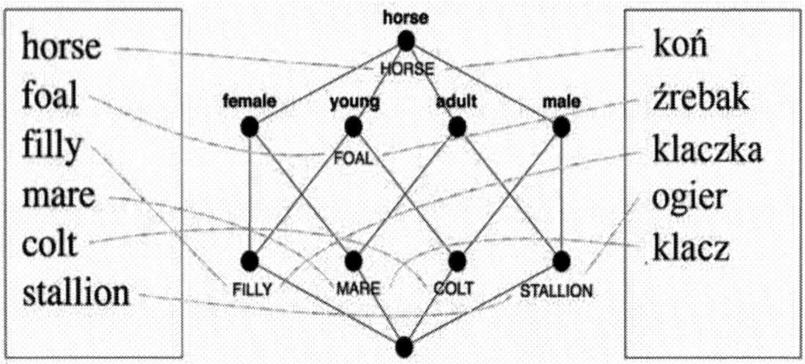

Figure 4-35 The concept lattice for horses after Janssen (2004, p. 145)

The concept lattice consists of nodes and links between these nodes. Nodes are assigned to every item in Table 4-21. The lattice is hierarchical. The top concept is *horse* and the nodes directly under the horse are definitional attributes. All the nodes under these nodes are concepts delineating different types of horses. Links connect definitional attributes with concepts, e.g. **horse**+**female**+**adult**=mare (in WordNet definitional attributes are in blue underline). The box on the left provides lexicalised horse concepts in English and the box to the right of the lattice gives lexicalised horse names in Polish. These corresponding names in English and Polish are connected through the relevant concept. The lattice eases the recognition of lexical gaps as, following the grey line, it may be spotted that there is no Polish word connected to the interlingual meaning (concept) of COLT. *Colt* has no translational synonym in Polish and is therefore classified as a conceptual gap. *Źrebak* is an approximate translation of *colt* but it does not have a definitional attribute **male**.

SIMuLLDA offers a solution to the translation of such lexical gaps. It is sufficient to study definitional attributes of colt in the lattice. It may be noticed that COLT shares one attribute with FOAL, that is **young**. In fact, FOAL is just a **young horse**, while COLT is defined as **young male horse**. What is more, there is a Polish word connected to the concept of FOAL, which is *ŹREBAK*. Thus, the complete meaning of COLT is FOAL+**male**. The translation in Polish may be found by giving the lexicalisation in Polish for these two components. The lexicalisation of FOAL in Polish is źrebak, and lexicalisation of **male** is ***rodzaj męski*** which means that a complete translation of COLT is *ŹREBAK* ***rodzaj męskiego***. It is also important to pay attention to the form *klaczka* 'filly' in Polish, where the ending *-ka* indicates that it is a diminutive form of *klacz* 'mare'.

Janssen's solution to the identification of conceptual mismatches based on componential analysis seems to be easier to follow and more transparent in rendering the results than the infon lattice by Kameyama et al. Therefore, it may be worth applying when the concept systems of two languages need to be compared and mismatches have to be identified. The strategy offered by Janssen to solve conceptual mismatches complements strategies mentioned by Kameyama et al. (1993) and Prahl and Pretzolt (1997) which include generalisation and specialisation. His hyperonymic approach with structured interlingua solves a lexical gap by providing a lexicalised hyperonym with a definitional attribute that distinguishes a given concept from related concepts. The fact that results can be plotted visually to assist in making an overall comparison is a benefit of this strategy.

The importance of conceptual analysis as a method of dealing with conceptual mismatches is also recognised in translation. Newmark (1988, p. 114) recognises it as one of the possible solutions which may be applied when a translation problem occurs. At the same time, he identifies the different roles of componential analysis in linguistics and translation. Componential analysis in linguistics depends on dividing various senses of a word into sense components, whereas, in translation, it compares a SL word with a TL word with a similar meaning; which is not an obvious one-to-one equivalent. It is achieved by indicating first the sense components they have in common and then their differing semantic features. Usually, the meaning of the SL word is more specific than the meaning of the TL word and, to produce a close approximation of the SL word meaning, a translator has to add a few TL features to the corresponding target language word as in (30).

(30) *toddler* = *dziecko* 'child' (+ *małe* 'young' + *które uczy się chodzić lub niedawno nauczyło się chodzić* 'who is learning or has recently learned to walk')

Linguistics uses tree diagrams (for single words), matrix diagrams or scalar diagrams to present componential analyses, while translation applies mainly equations as in (30) (Newmark, 1988, p. 115).

Componential analysis in translation may sometimes use matrix diagrams for source language lexical sets and scalar diagrams for lexical series. By a lexical set Newmark (1988, p. 121) understands an unordered set of objects, e.g. different types of bread, while a series covers a group of objects that are in some order, e.g. local government administrative units of different levels, military ranks. An example of a matrix diagram is

Table 4-21 above, which differentiates different types of horses based on such semantic components as male, female, adult and young. In this method, the semantic component is either present or not in the analysis. A scalar diagram offers a different method of presentation, as it does not focus on the semantic components, but rather tries to establish where the boundaries between the terms in the two languages lie. Colour terms, which do not have equal distribution across languages may be very well-illustrated using this method. For example, Welsh divides the green-brown part of the spectrum quite differently from English (Figure 4-36).

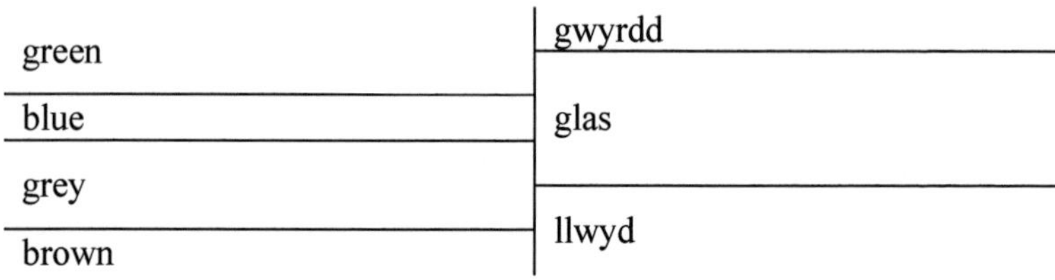

Figure 4-36 Scalar diagram for colours in English and Welsh (Crystal, 2008, p. 110)

The scalar diagram in Figure 4-36 illustrates different patterns of the use of colour words in English and Welsh. The word *glas* in Welsh is used for the colour of growing things. It does not correspond to one particular colour in English, as when the reference is made to growing things it may include blue, green and grey. The colour spectrum is a continuous band and it lacks any clear boundaries (Crystal, 2008, p. 110).

Green, blue, grey and brown belong to a group of eleven basic colours specified by Berlin and Kay (1969, p. 6). In addition to them, the group includes white, black, red, yellow, purple, pink and orange. Berlin and Kay are universalists and claim that all speech communities make use of basically the same set of colour terms. However Wierzbicka, who is also considered a universalist (Wyler, 1992, p. 13) postulates a twelfth basic colour term for Polish: *granatowy* 'black and blue', and three semi-basic colour terms: *beżowy* 'beige', *kremowy* 'off-white' and *bordowy* 'maroon' (1990, p. 138). Therefore, she proves that colour perception is not universal but culture-dependant.

Newmark (1988, p. 122) highlights the wide range of applications of componential analysis. He points out that it can be used for translating cultural and institutional words where it would require providing at least one descriptive and one functional component, e.g. *gmina* 'the principal unit of territorial division in Poland at its lowest uniform level'. It is also useful in translating neologisms, as it offers solutions for dealing with new words that name newly invented or imported objects and properties.

Componential analysis is, in fact, one of the main translation procedures proposed by Newmark (1988, p. 81). By translation procedures, he understands those solutions which are used for sentences and the smaller units of language, and he differentiates them from translation methods, which relate to whole texts. As I am dealing with terms, I will focus on translation procedures rather than translation methods. Furthermore, in my discussion of Newmark's translation procedures, I will only consider those techniques of translation that refer to terms and I will not tackle the procedures that are applied to sentences. Newmark proposes the following translation procedures that can be applied to terms:

- *transference* (emprunt, loan words, adoption, transfer), which depends on transferring a SL word to a TL text. It includes transliteration, which relates to the conversion of different alphabets. The word then becomes a loan word, e.g. the word *samovar* in English was transferred from Russian *самовар* <samovar>. The transference procedure is applied to transfer cultural objects. Brand names, names of private companies and institutions, names of public or nationalised institutions are also transcribed;
- *naturalisation*, which is the procedure that follows transference and adapts the SL word first to the pronunciation, and then to the morphology of the TL, e.g. *geokodowanie*[3] in Polish (from English *geocoding*);
- *cultural equivalence*, which is an approximate translation where a cultural word in the SL is translated by a cultural word in the TL, e.g. English *prom* is translated as *studniówka* in Polish 'graduation ball that takes place approx. 100 days before A-level exams';
- *functional equivalent*, which requires the use of a culture-free word and therefore neutralises or generalises the SL word and sometimes adds information, e.g. *GCSE exam* is translated into Polish as *egzamin zdawany w Wielkiej Brytanii (poza Szkocją) przez młodzież w wieku szesnastu lat* 'exam taken in the UK (excluding Scotland) by young people at the age of 16 years'. This procedure, which deculturalises a SL word, is a cultural componential analysis and comprises the most accurate way of translating. In the

[3] All the examples contrasting English and Polish are my own. Other ones were taken from Newmark (1988)

functional equivalence procedure, the focus is on the function of the SL word, which has to be conveyed when translating the word to the TL;

- *descriptive equivalent*, which is similar to the functional equivalent with the difference that description outweighs function, e.g. *cream tea* is described in Polish as *podwieczorek, na który składają się herbata, słodkie bułeczki, dżem i gęsta śmietana* 'afternoon snack including tea, sweet buns, jam and thick cream', wheras *high tea* is defined as *posiłek spożywany wczesnym wieczorem, na który składa się ciepłe danie takie jak ryba z frytkami, zapiekanka mięsno-warzywna lub makaron z serem, i po którym zjada się ciastka lub chleb z dżemem i masłem* 'an early evening meal, typically consisting of a warm dish such as fish and chips, shepherd's pie, or macaroni cheese, followed by cakes and bread, butter and jam';
- *synonymy*, which is translation of the SL word by a close TL equivalent. It is used when there is no clear one-to-one equivalent and when there are differences in meaning, e.g. English *quark* can be translated as *kostka sernikowa* 'a block of white cheese', i.e. smooth cottage cheese for a cheese cake. It is important to note that *quark* was borrowed into English from German where *der Quark* (also written as quarc, quarg, twarc) refers to a type of white cheese made of sour milk. The word has its origins in Slavic languages and it may come from Polish *twaróg* (Drosdowski, 1989, p. 1201).
- *through-translation* (calque, loan-translation), which is the literal translation of common collocations, names of organisations, the components of compounds. It is widely used for international institutional names, e.g. *Międzynarodowa Służba Geodynamiczna* 'International Geodynamic Service'. Many international organisations have acronyms which may be as widely used as the full names. They may remain untranslated in the SL and become internationalisms, e.g. *UNESCO*, or they may change in various languages, e.g. the English *EU* (European Union) is the Polish *UE* (Unia Europejska). Newmark claims that through-translations should be used only when they are already recognised terms;
- *shift* (Catford, 1965) or *transposition* (Vinay and Darbelnet, 1995), which involves a change in the grammar from SL to TL. This change with respect to terminology may involve a change from singular to plural, e.g. *furniture* is singular in English and its Polish equivalent *meble* is plural. Polish has a singular form for *furniture*, which is *mebel*, translated into English as *piece of furniture*;

- *recognised translation*, which is the official or the generally accepted translation of institutional terms, e.g. *Federal Aviation Administration* is translated into Polish as *Federalna Administracja Lotnicza*;
- *translation label*, which is a provisional translation usually applied to new institutional terms. It should be given in inverted commas when it first occurs in the text and without inverted commas when it recurs. The translation label is created through literal translation, e.g. *Computer Emergency Response Team (CERT)* translated into Polish as *Zespół Reagowania na Zagrożenia w Sieci* 'team for responding to threats on the web';
- *reduction* or *expansion*, e.g. translating *the Mesozoic era* into Polish as *mezozoik* is an example of reduction;
- *paraphrase*, which is an explanation of the meaning of a segment of text. It may be used to transfer the definition of the term in the SL to the TL if there is no equivalent for the source language term in the target language;
- *translation couplets*, *triplets*, *quadruplets*, which combine two, three or four of the above-mentioned procedures respectively to solve a single translation problem;
- *notes*, which include additional information on terms such as differences between SL and TL culture or the way in which the term is used (e.g. in standard language in ST and colloquial language in TT).

Apart from the general discussion of translation procedures, Newmark (1988, p. 32) approaches neologisms separately as he recognised them as one of the greatest challenges in translation. He specifies different types of neologisms and suggests how they should be dealt with. These are:

- old words with new senses, e.g. *web,* which tend to be non-cultural and non-technical. It used to refer to a spider's web only, but is currently applied to the Internet (world wide web). They are usually translated by a word that already exists in the TL or by a brief functional or descriptive term. *Web* in its modern sense referring to internet is translated into Polish as *sieć* 'web, net';
- new coinages, which include mainly brand or trade names, e.g. *Persil*, *Bacardi*. These are usually transferred, provided that the product is not marketed under a different name in the TL country. If the trade name does not have cultural or identifying significance,

the name of the product may be replaced by a generic or functional term;

- eponyms, which are usually left in their original form, although they may sometimes be transcribed, e.g. *wernier* in Polish (*vernier* in English) comes from Pierre Vernier, the French inventor of this device;
- derived neologisms, which are forms with productive prefixes (i.e. *de-*, *mis-*, *non-*, *pre-*, *pro-*) and suffixes (e.g. *-ism*, *-ise*, *-isation*), e.g. *misdefine*, *encyclopaedism*;
- new collocations, for which different translation procedures can be used depending on the type of collocation. Literal translation may be used if a term is transparent, e.g. *acid rain* or a paraphrase - if the concept is not known in the TL culture, e.g. *working tax credit* is translated into Polish as *dodatek do dochodów dla osób pracujących* (supplement to income for working people). Each collocation is different and its translation has to take into account the context where it appears, the readership of the text for which it is designed, etc.;
- phrasal neologisms, e.g. *trade-off*, *work-out*. They are translated by their semantic equivalents, i.e. *kompromis* 'compromise', *trening* 'training';
- acronyms. International acronyms are usually through-translated, e.g. *EU* (European Union) and have their equivalents in other languages, *UE* (Unia Europejska). National acronyms are usually retained with, if necessary, a 'translation' of their function, rather than a meaning, e.g. *Électricité de France - EDF*, *the French Electricity Authority*. Some acronyms become internationalisms, e.g. *laser*, *UNESCO*;
- abbreviations, e.g. *uni*, *prof.* They are normalised (i.e. translated unabbreviated), unless there is a recognised equivalent;
- cultural words referring to food, e.g. *gnocchi*, clothes, e.g. *sari*, processes, e.g. *tandoori*, cultural manifestations, e.g. *kung fu*. They are usually transferred.

Baker (1992, p. 26-42) offers a similar, although less specific, set of translation strategies that may be applied at word level. These include:

- translation by a more general word (superordinate);
- translation by a more neutral/less expressive word;
- translation by a cultural substitution, which depends on replacing a culture-specific item with the target-language item, which has a

different propositional meaning, but a similar impact on the reader, e.g. *Robin Hood* is replaced with *Janosik (a Polish equivalent of Robin Hood);*

- translation using a loan word or loan word plus explanation, which is particularly useful for translation of culture-specific terms and modern concepts, e.g. *układ SI* 'SI system' (International System of units);
- translation by a paraphrase using a related word, e.g. *overlooking* may be translated as *położony nad* 'situated by';
- translation by a paraphrase using unrelated words, when an item expressed in the source language is not lexicalised at all in the target language. In such a case, translation may be provided by giving a hyperonym or unpacking the meaning of the source item.

Most of these strategies overlap with the solutions Newmark (1988) offers. I find Baker's translation by a cultural equivalent particularly useful for my purpose, as many national institutions have corresponding institutions in other countries and their names may be provided in the notes field in the termbase. Translation using a loan word plus an explanation may well be applied when translating international organisation names which are better known by their acronyms than by their full names, *Long Range Aid to Navigation (LORAN)* is translated into Polish as *system LORAN.*

When discussing translation strategies, skopos theory needs to be considered. The theory was developed by Hans Vermeer in Germany in the late 1970s. It indicated a shift from linguistic and formal theories to functional and sociocultural approach to translation. Skopos theory was inspired by communicative theory, action theory, text linguistics, text theory and movements in literary studies towards reception theories. It stressed that the translation of scientific texts, manuals, tourist guides, contracts, etc. requires taking into account contextual factors surrounding the translation. These factors comprise the culture of the intended readers of the target text and of the client who commissioned the translation, and, in particular, the function the text is supposed to perform in that culture and for those readers (Schäffner, 2009: 235). This prospective function of the text is the purpose for which the translation is performed. The word *skopos*, derived from Greek, is used as the technical term for the purpose of a translation.

Vermeer (1978: 100) claims that skopos must be defined before the translation begins as it determines translation methods and strategies that will be used to translate the text The translation process does not depend on the source text as such but on the skopos of the target text determined by client's needs. The skopos of the target text and of the source text may be different as it varies with text receivers. If the target text receivers are indeed different from the source text readers, the function of the text changes and the standard for the translation is no longer intertextual coherence with the source text but adequacy or appropriateness to the skopos, which determines the selection and arrangement of the content in the target text.

Skopos theory is very relevant to the translation of terms that have a legal dimension, such as the system of route classification in the UK and Poland, where the legal aspects of the source text in English and the most obvious tourist dimension of the target text in Polish as well as hypothetical legal side of the target text needs to be considered. Skopos theory is also applicable to the translation of land registration terms as, depending on the function of the target text (general vs specific text) and its intended readers, different translation strategies will be used.

Apart from translation strategies that have already been discussed, some attention needs to be paid to ontologies and their potential in identifying conceptual mismatches. I have made an attempt to create an ontology for geodetic surveying equipment using CAOS. The process of creating an ontology is, in some respects, similar to componential analysis, as it requires analysing and predicting all types of relations that link concepts of a given field. Unlike componential analysis, however, CAOS ontology is illustrated with a tree diagram. Concepts within the tree are modelled by means of boxes. The tree diagram illustrates type relations (hyperonymy - hyponymy). The text boxes placed on the arrows within the tree structure indicate criteria on the basis of which hyperonym-hyponym relations were established. These criteria are repeated in each concept box and play the role of attributes, whose values are unique on a given level within the tree structure and facilitate making a distinction between a given concept and its co-hyponyms.

4.5 Classification and solution of translation problems resulting from conceptual mismatches

In this section, I look at entries in my termbases that pose translation problems and try to solve them by applying the translation strategies presented in section 4.4. I look at each case described in section 4.3 individually and then try to classify these problems and then attempt to develop translation solutions that could be applied to the remaining set of translation problems that have not been discussed in such detail in section 4.3.

Translation problems are quite diverse as, apart from conceptual mismatches that constitute gaps in the concept systems, there are also national institutional names that are country-specific and require the application of a different set of translation strategies. I have adopted the convention of signposting conceptual mismatches with an asterisk before the equivalent in my termbases.

4.5.1 Case study 1: transit vs teodolit reiteracyjny

The analysis of the concept systems for *transits* and *theodolites* in English and Polish allowed me to establish that *transit* is a problematic term in the English termbase, while in the Polish termbase *teodolit reiteracyjny* is a term for which no English equivalent can be found. In the analysis of the term *transit*, I have to take into account that, depending on context, the term can have one of three meanings as shown in Table 4-22.

Table 4-22 Equivalents for *transit* in Polish

English definition	translation label	paraphrase of the English definition
astronomical instrument used to observe the passage (transit) of stars across any portion of the celestial meridian		instrument do obserwacji przejścia ciała niebieskiego (tranzytu) przez południk w połowie drogi między swoim wschodem a zachodem
surveying instrument of lower quality than theodolite which has an "open-circle" design which allows an operator to see its graduated circle and read it with the aid of verniers	'tranzyt geodezyjny'	instrument geodezyjny o niższej dokładności niż teodolit, w którym urządzenia odczytowe nie są obudowane dzięki czemu jest możliwy bezpośredni odczyt pomierzonego kąta z tych urządzeń przy pomocy noniusza
theodolite whose telescope could be transited or reversed by rotating it about a horizontal axis (transiting theodolite)		teodolit z lunetą przechylaną przez zenit

Although all the senses of the term *transit* presented in Table 4-22 are known in Polish, only the first sense is lexicalised in Polish as *koło południkowe* 'longitudinal circle'. The second sense may be translated into Polish by providing a translation label, which indicates in what field an instrument is used. Thus, *tranzyt geodezyjny* consists of the term *tranzyt* 'transit' and the relational adjective *geodezyjny* 'geodetic' indicating that it is used in the geodetic field. The term *transit* remains ambiguous for the Polish user, and for this reason translation labels are accompanied by paraphrases of English definitions which are provided in the *Notes* field in the termbase. The third sense of the term may be translated into Polish by paraphrasing an English definition, which is concise and explicit at the same time.

Teodolit reiteracyjny in Polish seems to pose similar problems in translation. By looking at the hierarchical structure of the concept systems of single- and double-axis theodolites and definitions of different types of theodolites, I managed to establish that the Polish terms *teodolit repetycyjny* and *teodolit z układem jednoosiowym* 'single-axis theodolite' correspond to *repeating theodolite* and *directional theodolite* in English respectively, while the term *teodolit reiteracyjny* 'reiterating theodolite' does not have a lexicalised equivalent in English. It can be translated by using a hyperonym and a descriptive equivalent as 'theodolite equipped with a mechanism that enables it to rotate the limb independently from the alidade'.

4.5.2 Case study 2: levels

A detailed analysis of the concept systems for levelling instruments and their parts in English and Polish indicates that the systems of the two languages have many common points. Despite the fact that various organisations and authors use term labels inconsistently and the same name for holonyms and meronyms, the majority of English concepts have lexicalised equivalents in Polish, as concepts exist in the two languages. English terms and their Polish equivalents are presented in Table 4-23.

Table 4-23 English terms based on *level* referring to levelling instruments and their parts along with their Polish equivalents

English term	Polish equivalent
level	niwelator/libela
spirit level	niwelator libelowy/libela
compensator level	niwelator automatyczny
digital level	niwelator cyfrowy
optical level	niwelator optyczny
Abney level	pochyłościomierz, klizymetr
circular level	libela pudełkowa
tubular level	libela rurkowa

The terms *level* and *spirit level* pose translation problems as a translator must be aware of the fact that they have two meanings. They refer both to instruments and instrument parts. Each term has two equivalents in Polish: one for the instruments and one for the instrument components.

Equivalents for the English terms *dumpy level*, *tilting level* and *wye level* were not encountered in the Polish terminology. Thus, these terms are classified as conceptual mismatches. They are quite archaic. It is quite likely that they may have not been very popular in Poland and, therefore, their names have not been lexicalised. Definitions of these terms along with their equivalents in Polish are rendered inTable 4-24.

Table 4-24 Polish equivalents for *dumpy*, *tilting* and *wye level*

English term	definition	Polish equivalent
dumpy level	level in which the telescope is permanently attached to the base carrying the spirit levels, either rigidly or by a hinge about which the telescope can be rotated by means of a micrometer screw piece	niwelator z lunetą stałą 'level with a permanent telescope'
tilting level	levelling instrument in which the line of sight is brought into its final, level position by rotating the telescope on its trunnions	niwelator z przekładaną lunetą 'level with a rotating telescope'
wye level	level whose telescope rests in supports on the level bar called wyes	niwelator z lunetą mocowaną przy pomocy pierścieni w kształcie liery Y 'level with a telescope mounted using Y-shape rings'

The translation strategy applied for *dumpy* and *tilting levels* involves generalisation as only a limited part of the information included in the source language definitions is included in the target text equivalents. The motivation for this approach is the use of such terms in Polish by the *Muzeum Techniki Drogowej i Mostowej* ('Museum of Technology for Roads and Bridges') of the *Generalna Dyrekcja Dróg Krajowych i Autostrad* ('The Main Management of the state Roads and Motorways') (2010) The equivalent of *wye level*, not encountered in the Polish literature, is a translation of its definition.

4.5.3 Case study 3: surveying vs geodesy

The name of the field is not free from conceptual mismatches. Problems with naming the field have already been broadly discussed in (1.2) and (4.3.3). The hierarchical structures of the English and Polish concept systems are quite different. The Polish system, which specifies as many as eleven subfields of geodesy, seems to be more transparent than the English system, where different sets of surveying fields are specified but which are not correlated. Nevertheless, many matches between the two concept systems can be established after the analysis of definitions. They are presented in Table 4-25.

Table 4-25 Matching English and Polish surveying terms

surveying	geodezja
geomatics	geomatyka
geodesy	geodezja
geodetic surveying	geodezja wyższa
plane surveying	geodezja ogólna, geodezja niższa, miernictwo geodezyjne

There is a concept *of civil engineering*, which has a matching concept in Polish; that of *budownictwo lądowe i wodne* 'land and water building'. The concept has two hyponyms: *engineering surveying* and *geospatial surveying*. The former corresponds to the existing concept of *geodezja inżynieryjna* in Polish, while the latter shares many features with *Geomatics* and can be understood as *geomatyka* in Polish.

The English term *land surveying*, which was the name of the field at the very beginning, as surveys were limited to measurement of boundary lines, nowadays refers to one of the activities which are a part of surveying; as do its hyponyms: *boundary surveying* and *cadastral surveying*. Therefore, based on their definition and their current role, *land surveys* may be translated as *pomiary sytuacyjne* 'measurements of situation' (Newmark, 1988), *boundary surveying* as *pomiary granic działek gruntowych* 'measurements of borders of parcels of land' (Szejba et al., 1983) and *cadastral surveying* as *pomiary katastralne* (Hycner & Dobrowolska-Wesołowska, 2008). The equivalent for *land survey* was found by componential analysis, which involved studying definitions of these types of surveys in English and examination of the whole set of surveys in Poland. There are slight differences between *land survey* and *pomiar sytuacyjny* as the latter is concerned both with measurements of borders of the parcels, buildings and other objects that are part of the parcel and which need to be measured for the cadastral purposes, and also focuses on measuring such objects as bridges, railways, tunnels, monuments, kerbs, lamps, etc.

There are also three hyponyms in the Polish concept system which are not lexicalised in English. They are signposted in the termbase by an asterisk in the equivalent field. These are: *geodezja gospodarcza* 'economic surveying', *geodezja dynamiczna* 'dynamic surveying' and *instrumentoznawstwo geodezyjne* 'knowledge of surveying instruments'. Their English equivalents may be provided by translation labels that are direct translations which are accompanied by translations of their definitions as: *branch of surveying that deals with projects in administration, industry, communication, agriculture, forestry, mining and rails*; *geodetic surveying which applies gravimetric measurements*, and *branch of surveying concerned with construction, examination, usage and maintenance of surveying equipment*. Thus, in the termbase in the equivalent field translation labels are given, whereas the note field includes translations of their definitions as given above. When such terms occur in the running text, their first occurrences are translated by translation labels and paraphrases, and each subsequent occurrence just by translation labels.

Not only is the name of the field a challenge for a translator, so is the name of the profession. In principle, *surveyor* has a direct equivalent in Polish, which is *geodeta*. In the UK, a surveyor becomes a chartered surveyor (Polish equivalent – *geodeta uprawniony*) if he/she graduates from a program that is accredited by the Royal Institution of Chartered Surveyors and passes the Assessment of Professional Competence (APC), which is a two-year structured, work-based training scheme. Then, the surveyor gets a specialisation in some field and becomes one of the following (Goraj, Garliński & Przybyłowski, 1992):

- *Chartered Quantity Surveyor - geodeta kosztorysujący;*
- *Chartered Planning and Development Surveyor - geodeta do spaw planowania i rozwoju;*
- *Chartered Building Surveyor - geodeta budowlany;*
- Chartered Land and Hydrographic Surveyor - geodeta pomiarów ziemi i kartowania hydrograficznego;
- Chartered Mineral Surveyor – geodeta górniczy;
- *Chartered Land Agency and Agriculture Surveyor – geodeta ds. rolnictwa i zarządzania ziemią.*

In Polish, the majority of these concepts are not lexicalised as the focus is on the name of the chartership rather than on the name of the profession. In Poland, chartership may be obtained in the following areas of geodesy:

- *geodezyjne pomiary sytuacyjno-wysokościowe realizacyjne i inwentaryzacyjne* ('geodetic measurements of situation and elevations taking place during realisation and final control of the project'), i.e. plane and vertical land surveys during the project and after it has been completed;
- *rozgraniczanie, podziały i szacownie nieruchomości (gruntów) oraz sporządzanie dokumentacji do celów prawnych* ('establishing borders, partitions and evaluation of land estates and preparing documents for legal purposes'), i.e. border surveys and cadastral surveys and preparing cadastral documentation that has a legal value in Poland;
- *geodezyjne pomiary podstawowe*, ('geodetic measurements of basic level'), i.e. planimetric surveys;
- *geodezyjna obsługa inwestycji*, ('geodetic servicing of investments'), i.e. engineering surveying;
- *geodezyjne urządzenie terenów rolnych i leśnych*, ('geodetic organisation of terrains for agriculture and forests'), i.e. 'surveying of farm land and forests';
- *redakcja map*, ('editing of maps'), i.e. mapping;
- *fotogrametria i teledetekcja* ('photogrammetry and remote sensing').

As English names of surveying professions and Polish charterships do not overlap, they are translated into the target language by direct translations, which reflect the nature of qualifications held in a given profession or acquired through the chartership.

4.5.4 Case study 4: surveying assistant vs chainman

Polish has a single concept for *surveying assistant*, which is *pomiarowy*, while English has a wide range of terms such as *chainman*, *staff person*, *tape person*, *instrument man*, *umbrella man*, depending on the role the surveying assistant plays during the survey. Therefore, I may say that there are lexical gaps in Polish. These gaps can be filled relatively easily, however, as in the Polish termbase it is sufficient to generalise any of the English terms and refer to surveying assistant only. When any of these English terms occurs in a text that is translated into Polish, the translator will use the hyperonym and, when necessary, a definitional attribute that specifies the surveying assistant role. For example, *staff person* is translated into Polish as *asystent geodety trzymający łatę* 'surveying assistant who holds a staff'. The Polish term *pomiarowy* is translated into English as *surveying assistant*, as the direct equivalent exists.

4.5.5 Case study 5: aspect of projection and tangency

The English concepts *aspect of projection* and *tangency* were quite difficult to map across to Polish, as the Polish literature discusses how projections are arrived at, whereas the English literature focuses on the basic idea of projection. Thus, when translating the English concepts into Polish, I had to rely on materials that refer to cartographic visualisation. By analysing different sources on cartography, I managed to establish that these concepts are in fact lexicalised in Polish, although their names are used quite rarely. The criteria for distinguishing different types of projections in English have matching criteria in Polish. They may be found in Ogorzelska (2006, p. 84) and they include:

- *rodzaj powierzchni rzutu* ('type of projection surface');
- *położenie powierzchni rzutu w stosunku do bieguna kuli* ('location of the projection surface with respect to the sphere pole'), i.e. aspect of projection;
- *odległość powierzchni rzutu od kuli* ('distance of the projection surface from the sphere'), i.e. tangency or secancy;
- *położenie środka rzutu* ('location of the projection centre'), i.e. point of perspective.

In order to find translations of Polish criteria, I had to go to the field of mathematics, in particular geometry, to establish how they lead to creation of maps in different projections. There are no conceptual mismatches as far as the criteria are concerned. However, having analysed types of projection that were differentiated according to these criteria, I established that there is a concept *odwzorowanie odległe* 'distant projection' in Polish, which is not lexicalised in English, although the concept, as such, is known. It is due to the fact that the English system makes a distinction only between *tangent projection,* in which the projection surface touches the sphere, and *secant projection,* in which the projection surface cuts the sphere. The Polish concept system considers a third situation, when the projection surface does not touch the sphere, which is lexicalised as *odwzorowanie odległe.* The English equivalent can be created by loan translation as the expression *distant projection* is explicit in English.

4.5.6 Case study 6: mapping methods

The English and Polish concept systems for mapping methods differ in a few respects. The English cartographers refer to map types, while Polish cartographers focus on mapping methods that are used to create different types of maps. Furthermore, very few English authors address the problem of qualitative maps, as the main interest seems to lie in quantitative mapping. In Polish, by contrast, the two types of maps receive the same attention. English and Polish concepts referring to mapping methods were matched by comparing definitions and pictures illustrating different types of maps resulting from the application of the mapping techniques. The results of the matching process are shown inTable 4-26.

Table 4-26 Matching English and Polish terms referring to mapping methods

qualitative mapping methods	
nominal point/line symbol mapping	metoda sygnatur 'signature method'
R.S. land use mapping	metoda zasięgów 'method of ranges'
chorochromatic mapping	metoda chorochromatyczna 'chorochromatic method'
quantitative mapping methods	
choropleth mapping	kartogram
isarithmic mapping	metoda izolini 'isoline method'
dot mapping	metoda kropkowa 'dot method'
diagram mapping	metoda kartodiagramu 'diagram method'
proportional point symbol mapping	metoda sygnatur ilościowych 'method of signatures of quantity'
cartogram/value-by-area mapping	pseudokartogram 'pseudocartogram' or mapa anamorficzna 'anamorphic map'
flow mapping	kartodiagram liniowy 'line diagram'
statistical surface method	generowanie modelu 3D na podstawie mapy izolinii 'generating of 3D model on the basis of isoline map'

The process of finding equivalents between the two systems was quite straightforward, in the case of qualitative methods, as the number of concepts is the same. By analysing definitions, I could reassure myself that *chorochromatic method* in English corresponds to *metoda chorochromatyczna* in Polish, and *nominal point symbol mapping* and *metoda sygnatur* are equivalent. The concept of *R.S. land use mapping* is more tricky as the method presents areas of different types of land use which do not overlap, while *metoda zasięgów* 'range method', i.e. spatial reach method, which seems to be a corresponding concept, has a wider meaning as it presents any type of phenomena that do not overlap. Therefore, it is correct to translate *R.S. land use map* as 'metoda zasięgów', but translating *metoda zasięgów* as 'R.S. land use mapping' will be correct only if the reference is to land use. In other cases, the term should be translated by its descriptive equivalent as *spatial reach method.*

The area of quantitative mapping provides a challenge to the translator due to the fact that the English system has as many as eight different mapping methods, while the Polish system has only four. By comparing definitions and pictures of maps, I found out that *metoda izolinii* and *isarithmic mapping*, *metoda kropkowa* and *dot mapping* and *metoda kartodiagramu* and *diagram mapping* are matching concepts. The Polish *kartogram* and English *cartogram* turned out to be false friends as the Polish *kartogram* is used to present the intensity of a given phenomenon within the reference units using colour or shaded symbols, so it plays exactly the same role as *choropleth mapping* in English. The question is then, what is the equivalent for the English term *cartogram*? By consulting experts in the field of cartography and discussing the definition and pictures of cartograms with them, I managed to establish that Polish cartographers refer to them as *pseudokartogramy* 'pseudocartograms' or *mapy anamorficzne* 'anamorphic maps'.

Proportional symbol mapping, which is a concept in the English concept system, does not have a corresponding concept in the system of quantitative methods in Polish and is a conceptual mismatch. However, if it were included in the Polish concept system, it would be classified as a quantitative variation of the qualitative *metoda sygnatur* 'signature method', and could be translated into Polish as *metoda sygnatur ilościowych* 'method of signatures of quantity'.

The English concept system specifies flow mapping and statistical surface method as quantitative methods. These methods do not occur in the Polish concept system. It is a well-established fact that isoline maps may be transformed into 3D models; which are statistical surfaces. Thus, the concept of *statistical surfaces,* as such, is possible in Polish, but it is not lexicalised, while the concept of 3D model is lexicalised. Therefore, I would suggest using generalisation as the technique of providing the equivalent concept and I would translate the *statistical surface method* as *generowanie modelu 3D na podstawie mapy izolinii.*

The technique of *flow mapping* is not present in the Polish concept system either. However, by comparing the flow line map with different pictures of Polish diagram maps, I discovered that, in fact, *kartodiagram liniowy* 'line diagram' looks very much the same as the English flow map. The line diagram illustrates connections between areas or points and their intensity by the width of the line. The two maps serve the same purpose although the techniques used to create them are classified differently in the two languages. The reason for this may be the fact that, in the English concept system, a diagram is a deprecated form of cartographic presentation, and probably for this reason, the category of the flow map was introduced. In Polish, *metoda kartodiagramu* 'diagram mapping' is still listed as one of the main forms of presentation.

4.5.7 Case study 7: public rights of way

The English and Polish systems for public rights of way differ in many respects. The English concept system is based on legislation, while the Polish concept system does not have a very strong legal basis and is concerned mainly with environmental protection and is applied in tourism.

The English concept system is well developed, with formal definitions provided for different types of public rights of way, which include byways open to all traffic, restricted byways, bridleways and footpaths. It is very important to note that public rights of way in England and Wales are established on private lands when the piece of land is dedicated by its owner to public use. Public rights of way in the UK are recorded in a legal document called the Definitive Map and Statement. The map, drawn to a scale of 1: 10,000 or smaller shows public routes, while the statement contains particulars relating to the position and width of the path or any limitations or conditions affecting the right of way.

The Polish concept system specifies four different types of paths and trails (4.3.7): *ścieżki piesze* 'footpaths', *szlaki piesze* 'foot trails', *ścieżki i szlaki rowerowe* 'bike paths and trails' and *szlaki do jazdy konnej* 'horse trails'. There are no formal definitions of these routes, just a prototypical understanding of what they refer to. The Polish concept system does not have any concept that corresponds to *Definitive Map and Statement* but different maps of tourist trails are created to show paths and trails.

All these differences between the concept systems of the two languages result in conceptual mismatches. There are no straightforward solutions when translating from English to Polish or in the opposite direction. It is important to note that the problem of conceptual mismatches within the area of public rights of way may be approached from two different perspectives depending on the translation skopos. If the purpose of translation is to translate a tourist brochure, a translator does not have to look at the ownership of the land, but simply to find the nearest equivalent terms. When using this perspective, the English terms related to public rights of way may be translated as follows:

> *byway open to all traffic – szlak turystyczny* 'tourist trail';
> *restricted byway – szlak konno-rowerowy* 'horse and cycle trail';
> *bridleway - szlak do jazdy konnej* 'horse trail';
> *footpath - ścieżka piesza* 'footpath'.

The strategies involved in finding equivalent terms involve the omission of the legal side of these concepts and a focus on the way they are used by general public. Thus, only the tourist dimension of these concepts is considered. The translation strategies used in this case encompass cultural equivalence and omission. Correspondingly, the Polish terms referring to rights of way may be translated by through-translations as forms such as *cycle path* or *horse trail* are recognised in English. The concept of *Definitive Map* may be translated into Polish as *mapa szlaków turystycznych* 'map of tourist trails' and *Definitive Statement* may be translated as *spis szlaków turystycznych* 'register of tourist trails'. The equivalents are created by paraphrasing the English definitions of these concepts.

If the translation skopos is translating a legal contract, ownership of the land is a crucial aspect. In this case, English concepts may be translated into Polish by paraphrasing. Where English concepts do not have correspondences in Polish, the solution to explaining their meaning lies in finding an indirect link, such as a legal concept that exists in both languages. This indirect link, which facilitates the transition between the English and Polish concept systems, is the concept of *easement,* which is present in English and Polish.

A private right of way is an easement in English, which means that it is a right over land for the benefit of other land. The right must be attached to a particular piece of land and cannot be used by the public generally (Sydenham, 2001, p. 5). An easement may be established for the class of roads known as occupation roads, which are laid out for the use of the occupier of particular lands. For example, there may have been cottages for farm workers which have been sold off and the right of the owners to use the adjoining lane may amount to an easement in common with the other cottage owners. These roads will not be highways unless there is also use by the general public (Sydenham, 2001, p. 6).

The concept of *easement* is well established in the Polish legal tradition, where *służebność drogi koniecznej* ('easement of essential road'), grants a driveway access to the owner of a lot that has no street front and who is allowed to use a particular segment of a neighbour's land to gain access to the road. By adopting this concept and referring to the definitions, English concepts for public rights of way may be translated as follows:

- *byway open to all traffic – służebność gruntowa ustanowiona w celu wyznaczenia szlaku turystycznego (*'easement of private land established for the purpose of creating a trail for tourists, i.e. a route along which the public have a right of way for vehicular and all other kinds of traffic');
- *restricted byway – służebność gruntowa ustanowiona w celu wyznaczenia szlaku konno-rowerowego* ('easement of private land established for the purpose of creating a trail for horses and bikes, i.e. a route along which the public have a right of way on foot, on horseback or leading a horse and on a pedal cycle');
- *bridleway - służebność gruntowa ustanowiona w celu wyznaczenia szlaku konnego* ('easement of private land established for the purpose of creating a trail for horses, i.e. a route along which the public have a right of way on foot, on horseback or leading a horse');
- *footpath - służebność gruntowa ustanowiona w celu wyznaczenia szlaku pieszego* ('easement of private land established for the purpose of creating a foot trail, i.e. a route along which the public have a right of way on foot only').

The Polish equivalent for *Definitive Map and Statement* must be strongly connected with the English definition and it may be translated as follows: *Definitive Map – mapa zawierająca informacje o służebnościach gruntowych na gruntach prywatnych ustanowionych w celu wyznaczenia szlaków turystycznych* 'map containing information about easements of private land established for the purpose of designating tourist trails'; *Definitive statement – wykaz służebności gruntowych ustanowionych w celu wyznaczenia szlaków turystycznych* 'register of easements of private land established for the purpose of designating tourist trails'.

When translating Polish terms into English, direct translations may be provided as the paths and trails have only a tourist dimension and belong to the state.

4.5.8 Case study 8: land registration vs cadastre

The difficulties in finding equivalents for terms within the fields of cadastre and land registration are to some extent similar to the problems encountered when looking at correspondences in the previous subsection, where public rights of way were discussed. The differences in land law between countries result in conceptual mismatches within the domain of recording and registering rights to land. These conceptual mismatches are reflected in the name of the field as well as in its organisation, e.g. the different types of documents that are necessary to register rights to land.

Translation solutions offered to solve conceptual mismatches in this field will depend on the skopos of the particular translation job. If the target text is aimed to be a quite general text which only mentions land registration and is aimed at average users, it will be sufficient to translate any type of *land registration system* as *system katastralny* 'cadastral system', i.e. land registration system, thereby providing a cultural equivalent. The same principle can be applied when translating *system katastralny* into English.

However, if the target text is a specific and detailed text on land registration, the most explicit equivalents have to be provided. This may be achieved by looking at the definitions of concepts in the source language and providing their hyperonyms and descriptive and/or functional equivalents in the target language. Thus, the English land registration concepts may be translated into Polish as follows:

- *title registration system* as *system katastralny, w którym rejestruje się prawa do nieruchomości* 'cadastral system in which rights to the real estate are registered';
- *deed registration system* as *system katastralny, w którym rejestruje się transakcje dotyczące nieruchomości* 'cadastral system in which transactions concerning the real estate are registered';
- *private conveyance system* as *system katastralny, w którym spisuje się formalnie lub nieformalnie transakcje dotyczące nieruchomości* 'cadastral system in which transactions concerning the real estate are either formally or informally conveyed';
- *Torrens registration system* as *system katastralny wprowadzony przez Torrensa* 'cadastral system introduced by Torrens'.

When these terms occur in the text for the first time they are translated by providing translation labels (which are respectively: *system rejestracji tytułów do nieruchomości* 'system for registering titles to real estates', *system rejestracji transakcji* 'system for registering transactions', *system rejestrujący przeniesienie praw własności* 'system for registering conveyances', oraz *system Torrensa* 'Torrens system') accompanied by descriptive or functional equivalence. For any later reference, translation labels will be sufficient.

Documents, which are prepared when the title is registered, do not have equivalents in Polish. One such document called *title register,* can be translated into Polish as *rejestr praw własności i ograniczeń, które ciążą na tych prawach* 'register of ownership and limitations that affect it'. *Title register* is a *land register*, which includes both deeds and titles and may be translated into Polish as *rejestr gruntów.*

The second type of document, called *title plan,* may be translated as *plan pokazujący zakres występowania prawa własności do nieruchomości* 'plan which shows where the ownership of land stretches' or, in short, as *plan własności* 'plan of ownership'.

The Polish concept system of cadastre has three types of register which form the cadastre in Poland. They have recognised translations into English which are as follows:

- *ewidencja gruntów i budynków* - land and building register;
- *ewidencja podatkowa nieruchomości* - register of tax on properties;
- *księgi wieczyste* 'eternal books', i.e. land and mortgage register.

While the English system has the *title plan,* the Polish system uses *mapa katastralna,* which is translated into English as *cadastral map* (Tatarczyk, 2005). A cadastral map is similar to a title plan, but it also shows type of land-use and soil-based land classification.

Finding equivalents for Polish land registration units is also quite problematic as the UK has never had an integrated cadastral system and does not have such units specified. The lowest unit in the hierarchy of Polish cadastre is *działka ewidencyjna* 'parcel of land registration system', i.e. cadastral parcel, which may be translated as *land parcel,* as the two concepts are similar in English and Polish. The highest unit in the hierarchy is *jednostka ewidencyjna* 'cadastral unit', which may be translated as *basic land or property unit*, as it highlights the difference between the parcel and the unit. The concept *obręb ewidencyjny* 'cadastral precinct' is the most challenging one. It may be translated as *land precinct* between the parcel and the unit. It is worth using the notes field in the termbase to supply additional information on the structure of Polish cadastre, i.e. the Polish cadastre is organised at three levels: parcel level (the bottom level), precinct level (the intermediate level, which includes parcels within villages) and unit level (the top level, which includes all precincts within a town/city or district), whereas the British system is integrated.

4.5.9 General discussion

The case studies analysed above constitute a representative sample of the translation problems encountered in my termbases. I am not able to describe all individual examples in such detail, but on the strength of the data and evidence I have collected, I can match translation problems with translation strategies that can be used to tackle them.

Conceptual mismatches are the greatest challenge in translation as they require a componential analysis of the concept systems to which a problematic concept belongs, both in the source and in the target language. The componential analysis is followed by methods of filling lexical gaps, such as providing a hyperonym with a description and/or functional equivalent or paraphrasing. For example, the control networks in the UK and Poland are different, which is reflected in the fact that the concept *control network* has many hyponyms in Polish and only very few in English.

There also quite a few cases in the termbase when the SL concept is known in the target language, but it is not lexicalised. In such cases, translations are provided either by paraphrasing source language definitions and providing a hyperonym plus a definitional equivalent in the target language, e.g. *scene duration – zmienna wizualna przedstawiająca czas trwania zjawiska* 'visual variable that shows the duration of a given phenomenon', or by looking at how a concept is expressed in the target language, e.g. *map series*, which is a set of maps conforming to the same specifications, is translated to Polish as *podział mapy na arkusze* 'division of a map into map sheets'. The importance of the translation strategy which involves translation labels has to be highlighted. If a concept does not occur in the target language, it is introduced by providing a translation label, which is a direct translation of the source language term and, and followed by a paraphrase of the definition, or functional or descriptive equivalent, to provide the exact sense. If a term occurs in a source text a number of times, it is translated into a target language by providing a translation label and a paraphrase of its definition only when it is first encountered. For each subsequent occurrence, the translation label is sufficient.

The translation skopos is also a crucial factor to consider when the concept systems of two languages differ, as it suggests which translation strategy will be suitable. For example, when the target text is produced for tourist purposes, as in the case of the system of route classification, different route names are translated by direct translation. However, if the legal aspect is taken into account, translation has to be provided by paraphrasing the original definitions of terms in the target language.

The category of institutional names is also very interesting. National institutional names, such as *British Cartographic Society* or *Główny Urząd Geodezji i Kartografii* 'Main Surveying and Cartographic Office' are through-translated and their original names may follow the translation and may be provided in brackets or in inverted commas. Many of these institutions may have corresponding institutions in the target language country. It is useful to make a note of this fact in the notes field. The translation of English institutional terms is presented in Table 4-27.

Table 4-27 Translation of English national institutional terms into Polish

institution name	definition	translation into Polish	corresponding institution in Poland
British Cartographic Society	an organisation that associates individuals and organisations dedicated to exploring and developing the world of maps	Brytyjskie Towarzystwo Kartograficzne	Stowarzyszenie Kartografów Polskich 'Association of Polish Cartographers'
MasterMap	a database that contains a variety of information structured into different products	baza danych topograficznych Wielkiej Brytanii	Baza danych topograficznych
National Grid	a metric grid based on the Transverse Mercator Projection developed by Ordnance Survey in 1936 for use in Great Britain	państwowy system odniesień przestrzennych w Wielkiej Brytanii	Państwowy system odniesień przestrzennych

Ordnance Datum Newlyn	the national height system for mainland Great Britain which forms the reference frame for heights above mean sea level	państwowy system wysokości Newlyn	Państwowy system wysokości Kronsztadt 'Ordance Datum Kronsztadt'
Ordnance Survey	national mapping agency of the UK	państwowa służba kartograficzna w Wielkiej Brytanii	Centralny Ośrodek Dokumentacji Geodezyjnej i Kartograficznej 'Main Geodetic and Cartographic Documentation Centre', Główny Urząd Geodezji i Kartografii 'Main Surveying and Cartographic Office'
Royal Institution of Chartered Surveyors	the pre-eminent organisation for professionals working in the land, property and construction sectors in the UK and around the world	Brytyjski Królewski Instytut Dyplomowa-nych Rzeczoznaw-ców	Polskie Stowarzyszenie Rzeczoznaw-ców Wyceny nieruchomości 'Polish Association of Real Estate Appraisers'

The concept of *National Grid* exists both in English and in Polish but the grid is based on different coordinate systems and parameters in the two countries. The British system of surveying and mapping is more centralised than the Polish system, as the Ordnance Survey in the UK combines the duties of at least two Polish surveying organisations: *Ośrodek Dokumentacji Geodezyjnej i Kartograficznej* ('Main Geodetic and Cartographic Documentation Centre') and *Główny Urząd Geodezji i Kartografii* ('Main Surveying and Cartographic Office'). When national institution names are translated into a target language for the first time, direct translation is usually provided, followed by their English acronym. For each subsequent occurrence, the acronym is sufficient.

International institutional names comprise a separate category. Their translation is straightforward as they have recognised translations in a number of languages, e.g. *Global Positioning System – Globalny System Pozycjonowania, Federal Aviation Administration – Federalna Administracja Lotnicza.* It is important to note that their acronym is also transferred. For example, English *Land Information System (LIS)* is *System Informacji o Terenie (SIT) in* Polish. Some of the institutions are better known by their acronyms than by their full names. If they are translated into the target language, it is usually their acronym that is transferred. The TL equivalent is usually preceded by a hyperonym stating what type of entity a given term is, e.g. *International System of units* is translated into Polish as *układ SI*, *Long Range Aid to Navigation (LORAN)* has an equivalent *system LORAN* in Polish.

4.6 Summary

In this chapter, various theories of meaning were discussed and the mentalist theory was selected as the basis for the analysis of surveying concepts. The usefulness of ontologies for the organisation and representation of the concept systems was also elaborated on. The mentalist approach was applied to study conceptual mismatches in the domain of surveying, i.e. cases where there is no one-to-one correspondence between terms in English and Polish. The first indication of a conceptual mismatch is usually the lack of corresponding term in the target language. The methods that are used to confirm whether a given case is actually a conceptual mismatch involve componential analysis and ontologies. Conceptual mismatches often result in translation problems, which may be solved using various translation strategies. The case studies relating to conceptual mismatches, which have been discussed in depth in this chapter, together with all other examples of conceptual mismatches, have allowed me to identify the most efficient translation strategies for dealing with a particular type of conceptual mismatch. It has to be noted that I am discussing mainly concepts from three subfields of surveying here: geodetic surveying, cartography and GPS. The case studies referring to the system of routes and land registration may be considered as cadastral concepts. Whereas geodetic surveying, cartography and cadastre have established terms in the course of a more nationally-oriented history, GPS is a new field with many recently-added terms. I noticed that conceptual mismatches typically occur when the use of tools or methods is country- or culture-specific and they are not known in other places.This also happens when concepts relating to these methods are known but they do not have names in other languages, or when concepts in the source and the target language are similar, but not identical, as they are classified differently. It must be emphasised that GPS concepts are usually borrowed from English into Polish, often together with their English names or are not lexicalised in Polish and, therefore, direct translation is used to provide equivalents in Polish. Finding equivalents for terms representing concepts that have the 'legal dimension', such as systems of routes and land registration, requires taking translation skopos into account as these concepts are shared concepts occurring both in surveying and law.

The most common translation strategies used for dealing with conceptual mismatches are: translation by functional and descriptive equivalents, paraphrasing a definition, and specialisation and generalisation of meaning. The role of translation labels is also significant, especially if the concept occurs many times in the text. I also noticed that conceptual mismatches with a legal dimension may often be solved through other legal concepts, which are well established both in the source and the target language. For example, systems of routes may be explained in the context of easement, which is a well established concept in English and Polish.

CHAPTER FIVE

CONCLUSION

This study investigated terminological and conceptual differences in surveying terminology in English and Polish. It focuses primarily on three subfields of surveying: geodetic surveying, cartography and GPS. The analysis of terms and concepts within these domains proved that geodetic surveying and cartography developed quite independently in the UK and Poland as there are number of concepts that do not exist in one of the languages or concepts that are not lexicalised. Systems for organising human knowledge are culture-dependent, which is reflected in a different classification of concepts, e.g. mapping methods are categorised differently in English and Polish.

The field of GPS, which is one of most recent fields, has a well-established position in English-speaking countries, and plays a major role in the development of new equipment and technologies. Most of the material relating to this field is published in English, e.g. software and hardware manuals. Countries which are not directly engaged in the development of this technology, such as Poland, often borrow English concepts along with their names. Hence, there are many borrowings within this field in Polish.

The analysis of different case studies also confirmed that legal-related domains are particularly prone to conceptual mismatches. Legal systems differ between countries, especially if countries evolved independently and do not have many common points in history, as is the case of the UK and Poland. Conceptual mismatches were observed in cadastre and in the system of routes.

Finding equivalents is not always a straightforward task, as conceptual mismatches may often be solved in more than one way. Case studies examined in this research indicate that translation skopos is crucial in this process as it determines which translation strategy should be chosen in a particular context.

While making a significant contribution to surveying terminology the research also has a few limitations. The scope of the study is limited in two respects: one is the coverage of three subfields of surveying and the other is the corpus size. The research focuses on three out of the ten subfields of surveying specified in chapter one, which are geodetic surveying, cartography and GPS. The majority of entries in the termbases belong to these subfields. There are a few entries that belong to other fields, but there are not enough of such terms to get a full overview of the domain and relations between concepts.

The surveying corpora in English and Polish are relatively small which is reflected in the number of entries in the termbase. The English surveying corpus consists of 126,000 words, whereas the Polish one has approximately 116,000 words. Consequently, the English termbase includes 490 entries, whereas the Polish one contains 459 entries. The process of compiling corpora involved lots of manual work as the main sources of data were textbooks which were scanned and edited. Moreover, there was limited access to surveying sources in Polish, as, at the time when the Polish corpus was compiled (2008-2009), there were few surveying books available; some of which were outdated.

Although the scope of the research is limited, the design of the termbase makes it possible for it to be extended. The problematic cases I dealt with in detail in geodetic surveying, cartography and GPS in chapter four represent common problems that may occur in surveying terminology. I do not expect to find any further problems that cannot be solved using the same techniques.

The methodology that has been developed for compiling the termbases for the three subfields of surveying and for solving conceptual mismatches in these termbases can be used to extend the scope of the research in order to cover all ten subfields of surveying specified in (1.2.3) and to compile larger corpora which include more terms.

ABBREVIATIONS

ADJ	adjective
ADV	adverb
AGH	Akademia Górniczo-Hutnicza 'Academy of Mining and Metallurgy' i.e. AGH University of Science and Technology
AI	artificial intelligence
ANSI	American National Standards Institute
APC	Assessment of Professional Competence
ASCII	American Standard Code for Information Interchange
A+N (expression)	(expression consisting of) an adjective and a noun
CAOS	Computer Aided Ontology Structuring
CD-ROM	compact disk read only memory
DAT	dative
DTMS	dedicated terminology management system
DVD	digital versatile disc or digital video disc
EB	*Encyclopaedia Britannica*
EWN	EuroWordNet
f	feminine
GDP	Gross Domestic Product
GEN	genitive
GN	genitive adjective
GOLD	General Ontology for Linguistic Description
GPS	Global Positioning System
HMLR	Her Majesty Land Registry
HTML	HyperText Markup Language
ICM	Idealised Cognitive Model
IL	Interlingual Item
IMPERF	imperfective
INSTR	instrumental
IPI PAN	Instytut Podstaw Informatyki Polskiej Akademii Nauk 'Institute of Computer Science at the Polish Academy of Sciences'
ITD	Intermediate Translation Document
KWIC	Key Word in Context
LR	Land register
LU	lexical unit
m	masculine

MARTIF	Machine Readable Terminology Interchange Format
MDB	Microsoft Access database
MLLD	Multilingual Lexical Databases
ModE	Modern English
MWE	multi-word expressions
n	neuter
N	noun
NCF	neoclassical formative
n.d.	no date
NLP	Natural Language Processing
NP	noun phrase
N+GN (expression)	(expression consisting of) a noun in the nominative case and a noun in the genitive case
N+N (expression)	(expression consisting of) two nouns
N+RA (expression)	(expression consisting of) two a noun and a relational adjective
OE	Old English
OED	Oxford English Dictionary
OKBC	Open Knowledge Base Connectivity
OS	Ordnance Survey
OWL	Web Ontology Language
PDF	Portable Document Format
PERF	perfective
PP	prepositional phrase
PWN	Polskie Wydawnictwo Naukowe 'Polish Scientific Publishers'
QA	qualitative adjective
RA	relational adjective
RDF	Resource Description Framework
RICS	Royal Institution of Chartered Surveyors
RoS	Registers of Scotland
RTF	Rich Text Format
SFPW	*Słownik frekwencyjny polszczyzny współczesnej* 'A Frequency Dictionary of Contemporary Polish'
SGML	Standard Generalized Markup Language
SIMuLLDA	Structured Interlingua MultiLingual Lexical Database Application
SL	source language
ST	source text
sth	something
TDB	terminological data base
TE	translation equivalent
TL	target language
TMS	terminology management system
TMW	translator's workbench
TMX	Translation Memory Exchange Format

TXT	text file
UCAS	Universities and Colleges Admissions Services
UDC	Universal Decimal Classification
V	verb
VML	Vector Markup Language
XML	eXtensible Markup Language

GLOSSARY

acronym: an abbreviation formed of initial letters of multi-word units pronounced regularly, just like words, e.g. LIDAR (Light Detection and Ranging)
affix: a morpheme that is attached to a word stem to form a new word
alignment: the process whereby sections of a source text are linked up with their corresponding translations
annotation: the process of encoding additional information into a corpus. This information can be syntactic (e.g. part-of-speech tags) or semantic (e.g. distinguishing between different meanings of a word)
ASCII (American Standard Code for Information Interchange): a standard 7-bit character set developed by the American National Standard Institute (ANSI) that is used to represent 128 characters which include the Roman alphabet, Arabic numerals, and a selection of other symbols that appear on most keyboards (e.g., ! ? $, % &)
blending: the process of combining two (rarely three or more) words into one and deleting material from one or both of the source words, e.g. geomatics created by combining geography and mathematics
clipping: shortening of the word by omission of one or more syllables, e.g. flu (from influenza)
comparable corpus: a collection of individual, monolingual corpora that contain completely different texts in several languages
CAOS (Computer-Aided Ontology Structuring): a tool for constructing terminological ontologies, which are domain-specific and model concepts and the relations between them
bilingual concordance: a type of concordance that operates on an aligned parallel corpus by retrieving all the occurrences of a particular pattern and its immediate context
collocation: words that occur together in a greater than random probability (i.e. words that are often "found in each other's company")
compound: a word that consists of more than one stem
componential analysis: a type of definitional analysis which breaks meanings down into binary features (i.e. features with only two possible values, + for truth and - for false)
conceptual mismatch: a difference in the concept systems of the source language and the target language, resulting in a lexical discrepancy
concordance: an index to the words in a text
concordance: a tool used to produce a concordance
conversion: a type of derivation without any overt marking
corpus: a large collection of texts that have been compiled according to specific criteria

Dedicated Terminology Management System (DTMS): database management system that has been developed or configured specifically for the purpose of managing terminological data, e.g. MultiTerm, which is specifically oriented towards translation

derivation: the process of adding a prefix or suffix to an existing word in order to produce a new word

entity type: a type of ontological category that the conceptual constituent represents such as: EVENT, STATE, THING, PROPERTY, PLACE, PATH, TIME, AMOUNT

eXtensible Markup Language (XML): a markup language containing tabs or symbols that are used to describe data elements on a Web page

FrameNet: a computational lexicographic project whose purpose is to represent information about the semantic and syntactic properties of English words and encode this information in a database

geodesy: the science of measuring and monitoring the size and shape of the Earth and the location of points on its surface

GOLD (General Ontology for Linguistic Description): a specialised ontology which describes specialised concepts from the field of linguistics

holonymy: a semantic relation between a word denoting the whole and a word denoting a part of the whole, e.g. tree (holonym) and trunk (meronym)

homograph: a word that has the same spelling as another word, but is a different part of speech, e.g. cook and to cook

homonym: a word that has the same spelling and is the same part of speech as another word but has a different sense, e.g. bank (financial institution) and bank (of the river)

homophone: a word that has the same pronunciation as another word but is spelled differently and has a different meaning, e.g. mail and male

hyperonym: a semantic relation between a more general and a more specific word

Hypertext markup language (HTML): the markup language used to define the document display format for the World Wide Web

Idealised Cognitive Model (ICM): a complex structure consisting of units of understanding, by means of which people organise their knowledge

inflection: the process used to make a noun plural or to conjugate a verb

initialism: an abbreviation formed of initial letters of multi-word units pronounced by saying each individual letter e.g. GPS (Global Positioning System)

Key Word in Context (KWIC): a format of displaying concordance lines which puts the word-form under examination in the centre of each line, with left and right context

lemma: a particular form of the lexeme that is chosen by convention, e.g. measure is the lemma of the verb lexeme measure

lexeme: a set of forms taken by a single word, e.g. the English verb lexeme measure

lexical gap: a lexicalisation difference that occurs when a language expresses in a lexical unit what the other language expresses with a free combination of words

machine translation (MT): the process whereby a computer has the primary responsibility for the translation of a text

MARTIF (Machine Readable Terminology Interchange Format): a format which facilitates the interchange of terminological data among terminology management systems

monolingual concordance: a tool that operates on a monolingual corpus by retrieving all the occurrences of a particular search pattern and its immediate context and displaying these in an easy-to-read format such as KWIC

multi-word unit: a group of two or more words used to express a single concept

noise: items that are retrieved during the search but are not of interests (e.g. are not terms)

onomasiology: a specific semantic perspective which starts from concepts and looks for their names

ontology: specification of a conceptualisation used to help humans and computers share knowledge

paradigmatic relation: the relation between a set of linguistic items which, in some sense, constitute choices, so that only one of them may be present at a time in a given position

parallel corpus: bi- or multilingual corporus that contain original texts and their translations in one or more languages

precision: the capacity of the system to discriminate between those units which are terms and those which are not.

prefix: an affix added to the beginning of the word

recall: the capacity of the system to extract all terms from a document

Rich Text Format (RTF): a file format that allows the user to transfer formatted text and graphic from one word processor to another

semantics: the study of meaning

semiotics: the study that deals with social production of meaning from sign systems

semasiology: a specific semantic perspective which starts from forms and looks for their meaning

silence: items that are of interest (e.g. terms) but that are not retrieved during a search because the query was not well-formulated or comprehensive enough

Structured Interlingua MultiLingual Lexical Database Application (SIMuLLDA): a structured interlingua system; In SIMuLLDA every word of every language relates to as many interlingual meanings as it has senses and the interlingual meanings are related hierarchically

suffix: an affix added to the end of the word

surveying: the science, art and technology of determining the relative positions of points above, on, or beneath the Earth' s surface, or of establishing such points, and the presentation of this information either graphically or numerically

syntagmatic relation: the relation between any linguistic elements which are simultaneously present in a structure

tag: a label attached to a data element that contains information related to that element (e.g. information about how it should be displayed)

taxonomy: the science of biological classification, which is usually restricted to the classification of plants and animals

term extraction tool: a tool that attempt to analyse texts and automatically extract candidate terms

termbase: a collection of term records that can be searched electronically

terminology management system (TMS): a software application that allows users to create, store and, retrieve term records

token: an individual word in a corpus

translation label: provisional translation usually applied to new institutional terms

translation memory (TM): a type of linguistic database that is used to store and retrieve source texts and their translations so that translators can reuse segments of previous translations when they translate a new source text

type: a word form in a corpus, each instance of which is referred to as a token

Unicode: a superset of the ASCII character set that uses two bytes to encode each character set rather than one

Universal Decimal Classification (UDC): a system of library classification developed by the Belgian bibliographers Paul Otlet (1868-1944) and Henri la Fontaine (1854-1943) at the end of the 19th century

voice recognition: the processing of spoken words by a computer

wildcard: a character that can be used to represent one or more characters in a search string

word-frequency list: a list of the number of types and tokens contained in a corpus

WordNet: ontology for general language available for English developed under the direction of George A. Miller

BIBLIOGRAPHY

AGH University of Science and Technology (2011) 'Wydział Geodezji Górniczej i Inżynierii Środowiska'. [Online]. Available at: http://www.geod.agh.edu.pl/portal/index.php/strony/rekrutacja (Accessed: 22.10.2009).

Agirre, E., Aldezabal, I., Etxeberria, J., Izagirre, I., Mendizabal, K., Pociello, E. & Quintian, M. (2006) 'Improving the Basque Wordnet by Corpus Annotation', *Proceedings of Third International Wordnet Conference.* Jeju Island (South Korea).

Ahmad, K. & Rogers, M. (2001) 'Corpus Linguistics and Terminology Extraction ', in Wright, S.E. and Budin, G. (eds.) *Handbook of Terminology Management.* Amsterdam / Philadelphia: Benjamins.

Albin, J. (2003) 'Kierunki Modernizacji Polskiego Systemu Katastralnego', *Sprzątamy po ewidencji - przyszłość zawodu geodety* Pogorzelica. Available at: http://www.geodezja-szczecin.org.pl/stara_strona/ Konferencje/Konf2003/k01.html (Accessed: 10.10.2009).

Atkins, B. T. & Rundell, M. (2008) *The Oxford Guide to Practical Lexicography.* Oxford / New York: Oxford University Press.

Baker, C. F., Fillmore, C. J. & Cronin, B. (2003) 'The Structure of the Framenet Database', *International Journal of Lexicography*, 16 (3), pp. 281-296.

Baker, C. F. & Sato, H. (2003) 'The Framenet Data and Software'. [Poster and Demonstration at Association for Computational Linguistics, Sapporo, Japan].

Baker, M. (1992) *In Other Words: A Coursebook on Translation.* London: Routledge.

Bańko, M. (2005) *Wielki Słownik Wyrazów Obcych.* Warszawa: PWN.

Bannister, A., Raymond, S. & Baker, R. (1998) *Surveying.* 7th edn. Harlow: Longman.

Barkema, H. (1996) 'Idiomacity and Terminology: A Multi-Dimensional Descriptive Model', *Studia Linguistica*, 50 (2), pp. 125-160.

Bauer, L. (1983) *English Word-Formation.* Cambridge: Cambridge University Press

—. (2009) 'Typology of Compounds', in Lieber, R. and Stekauer, P. (eds.) *The Oxford Handbook of Compounding.* Oxford: Oxford University Press, pp. 343-356.

Beeken, J., Heid, U., Laureys, G., Martin, W. & Schuurman, I. (1998) 'On the Construction of Bilingual Dictionaries: Feasibility Study Carried out by Order of the European Commission DG XIII.'.[Technical Report] Stuttgart.

Benczes, R. (2010) 'Setting Limits on Creativity in the Production and Use of the Metaphorical and Metonymical Compounds', in Onysko, A. and Michel, S. (eds.) *Cognitive Perspectives on Word Formation.* Berlin / New York: Walter de Gruyter, pp. 219-242.

Bentivogli, L. & Pianta, E. (2000) 'Looking for Lexical Gaps', in Heid, U., Evert, S., Lehmann, E. and Rohrer, C. (eds.) *Proceedings of the Ninth Euralex International Congress.* Stuttgart: Institut für Maschinelle Sprachverarbeitung, Universität Stuttgart, pp. 663-669.

Berlin, B. & Kay, P. (1969) *Basic Color Terms: Their Universality and Evolution.* Berkeley: University of California Press.

Blank, A. (1999) 'Why Do New Meanings Occur? A Cognitive Typology of the Motivations for Lexical Semantic Change', in Blank, A. and Koch, P. (eds.) *Historical Semantics and Cognition (Cognitive Linguistic Research).* Berlin/New York: Mouton de Gruyter, pp. 61-90

Blank, A. (2001) 'Pathways of Lexicalization', in Haspelmath, M., König, E., Oesterreicher, W. and Raible, W. (eds.) *Language Typology and Language Universals.* Berlin: Walter de Gruyter, pp. 1596–1608.

Bloomfield, L. (1933) *Language.* London: George Allen & Unwin.

Booij, G. & van Marle, J. (2002) *Yearbook of Morphology 2001.* Dordrecht: Kluwer

Borst, W. (1997) *Construction of Engineering Ontologies for Knowledge Sharing and Reuse*. PhD thesis. University of Twente.

Boryś, W. (2005) *Słownik Etymologiczny Języka Polskiego.* Kraków: Wydawnictwo Literackie.

Boulanger, J. C. (1995) 'Présentation: Images et Parcours de la Socioterminologie', *Méta* XL (2), pp. 195-205.

Bowker, L. (2002) *Computer-Aided Translation Technology: A Practical Introduction - Didactics of Translation Series.* Ottawa: University of Ottawa Press.

Bowker, L. & Pearson, J. (2002) *Working with Specialized Language: A Practical Guide to Using Corpora* London/New York: Routledge.

Bowman, C., Michaud, D. & Suonuuti, H. (1997) 'Do's and Don'ts of Terminology Management ', in Wright, S.E. and Budin, G. (eds.) *Handbook of Terminology Management.* Amsterdam / Philadelphia: Benjamins, pp. 215-217.

British Standard Institution (1977) *BS 1000[52]:1977. Universal Decimal Classification. English Full Edition. Astronomy. Astrophysics. Space Research. Geodesy* London: BSI.

—. (2002) *Bs 7666-4:2002. Spatial Data-Sets for Geographical Referencing. Specification for Recording Public Rights of Way* London: BSI.

Bucewicz, B., Ciesielski, J., Januszko, W., Kowalczyk, A. & Umecki, R. (1981) Instrukcja Techniczna K-1. Mapa Zasadnicza. Warszawa: GUGiK.

Bureau of Land Management (2009) 'U.S. Department of the Interior. Bureau of Land Management'. [Online]. Available at: http://www.blm.gov/wo/st/en/prog/more/cadastralsurvey.html (Accessed: 26.04.2009).

Burger, H. (1998) *Phraseologie. Eine Einführung am Beispiel des Deutschen.* Berlin: Erich Schmidt.

Cabré, M. T. (1999) *Terminology: Theory, Methods and Applications.* Amsterdam / Philadelphia: Benjamins.

Cabré, M. T., Estopà, R. & Vivaldi, J. (2001) 'Automatic Term Detection: A Review of Current Systems', in Bourigault, D., Jacquemin, C. and L'Homme, M.-C. (eds.) *Recent Advances in Computational Terminology.* Amsterdam / Philadelphia: Benjamins, pp. 53-87.

Cambridge Advanced Learners' Dictionary (2011) 'Cambridge Dictionaries Online'. Cambridge University Press. [Online]. Available at: http://dictionary.cambridge.org/ (Accessed: 10.08.2011).

Campbell, J. (2000) *Map Use & Analysis.* Boston: McGraw-Hill.

Canada Centre for Remote Sensing (2005) 'Glossary of Remote Sensing Terms'. [Online]. Available at: http://www.ccrs.nrcan.gc.ca/glossary/index_e.php (Accessed: 18.02.2011).

Carter, W. E. (1965) *A Field Evaluation of the Kern' Dim3-a Astronomical Theodolite for Precise Astronomic Position Determination.* MSc Thesis. The Ohio State University.

Cartisan AGS (2010) 'Cartisan Maps'. [Online]. Available at: http://www.cartisan.com/portfolio/ (Accessed: 16.03.2011).

Catford, J. C. (1965) *A Linguistic Theory of Translation: An Essay in Applied Linguistics.* Oxford: Oxford University Press.

Centrum voor Vaktaal en Communicatie (2009) 'Centre for Special Language Studies and Communication'. [Online]. Available at: http://taalkunde.ehb.be/cvc (Accessed: 20.09.2009).

Chomsky, N. (1959) 'Verbal Behaviour', *Language*, 35 pp. 26-58.

Countryside Act 1968.

Countryside and Rights of Way Act 2000.

Cowie, A. P. (ed.) (1998) *Phraseology: Theory, Analysis, and Applications.* Oxford: Oxford University Press.

Cowie, A. P. (2009) *Semantics.* Oxford: Oxford University Press.

Crystal, D. (2008) *A Dictionary of Linguistics and Phonetics.* 6th edn. Malden MA, Oxford: Blackwell.

Cuff, D. J. & Mattson, M. T. (1982) *Thematic Maps : Their Design and Production* London: Methuen.

DeLoach, S., R (1994) 'Glossary of the Mapping Sciences'. New York: Joint Committee of the American Society for Photogrammetry and Remote Sensing, American Congress on Surveying and Mapping, American Society of Civil Engineers.

Dent, B. D. (1996) *Cartography. Thematic Map Design.* 4th edn. Dubuque, IA: Wm. C. Brown Publishers.

Derwojedowa, M., Piasecki, M., Szpakowicz, S. & Zawiławska, M. (2007) 'Polish Wordnet on a Shoestring', in Rehm, G., Witt, A. and Lemnitzer, L. (eds.) *Datenstrukturen Für Linguistische Ressourcen Und Ihre Anwendungen - Proceedings Der Gldvjahrestagung 2007.* Tübingen: Tübingen University, pp. 23-32.

Deumlich, F. (1982) *Surveying Instruments.* Berlin/New York: W. de Gruyter.

Di Sciullo, A. M. & Williams, E. (1987) *On the Definition of Word.* Cambridge, Mass.: MIT Press.

Dirven, R. & Verspoor, M. (2004) *Cognitive Exploration of Language and Linguistics (Cognitive Linguistics in Practice)* 2nd revised edn. Amsterdam / Philadelphia: Benjamins.

Downarowicz, J. & Leśniok, H. (2006) *Polsko-Angielski, Angielsko-Polski Słownik Terminów Z Zakresu Geodezji, Map I Nieruchomości.* Warszawa: Oficyna Wydawnicza Politechniki Warszawskiej.

Drosdowski, G. (1989) *Duden Deutsches Universal Wörterbuch.* Mannheim/Leipzig: Dudenverlag.

Dubisz, S. (2002) 'Słownictwo', in Dubisz, S. (ed.) *Nauka o Języku dla Polonistów.* 4th edn. Warszawa: Książka i Wiedza, pp. 327-385.

—. (2003) *Uniwersalny Slownik Języka Polskiego.* Warszawa: PWN.

Dubuc, R. & Lauriston, A. (1997) 'Terms and Contexts', in Wright, S.E. and Budin, G. (eds.) *Handbook of Terminology Management.* Amsterdam / Philadelphia: Benjamins, pp. 80-87.

Duggal, S. K. (2009) *Surveying.* 3rd edn. vol. 1. New Delhi: Tata McGraw-Hill Education.

Duhaime Organisation (n.d) 'Legal Dictionary'. [Online]. Available at: http://www.duhaime.org/LegalDictionary/T/TorrensLandRegistrationSystem.aspx (Accessed: 10.02.2011).

Emapa Transport (n.d) 'Mapy Cyfrowe dla Biznesu'. [Online]. Available at: http://www.e-mapa.pl/emapa_transport_spec.asp (Accessed: 22.10.2009).

Encyclopaedia Britannica (1986) 'The New Encyclopaedia Britannica in 32 Volumes', in Goetz, P.W. and Sutton, M. (eds.). 15th edn. London: Encyclopaedia Britannica (UK).

—. (2009) [CD-ROM]. Encyclopaedia Britannica.

European Comission. Joint Research Centre (2011) 'The Jrc-Acquis Multilingual Parallel Corpus'. [Online]. Available at: http://langtech.jrc.it/JRC-Acquis.html (Accessed: 6.09.2011).

Farvacque, C. & McAuslan, P. (1992) *Reforming Urban Land Policies and Institutions in Developing Countries.* Washington, D.C: The World Bank.

Fellbaum, C. (1990) 'English Verbs as a Semantic Net', *International Journal of Lexicography*, 3 (4), pp. 278-301.

—. (1998) *Wordnet: An Electronic Lexical Database* Cambridge, Mass.: MIT Press.

Fernández Parra, M. & ten Hacken, P. (2008) 'Multi-Word Units in Multiterm Extract', *Proceedings of the Thirtieth International Conference on Translating and the Computer, Aslib, 27-28 November 2008.* London.

Filip, H. (2008) 'What Is Semantics, What Is Meaning.'. [Unpublished Work] University of Florida.

Fillmore, C. (1975) *Santa Cruz Lectures on Deixis, 1971.* Bloomington: Indiana University Linguistics Club.

Fillmore, C. J. (1968) 'The Case for Case', in Bach, E. (ed.) *Universals in Linguistic Theory.* New York: Holt, Rinehart & Winston of Canada Ltd, pp. 1-88.

—. (1976) 'Frame Semantics and the Nature of Language', *Annals of the New York Academy of Sciences*, 280, pp. 20-32.

Fischer, R. (1998) *Lexical Change in Present-Day English : A Corpus-Based Study of the Motivation, Institutionalization, and Productivity of Creative Neologisms.* Tubingen: Gunter Narr.

FrameNet (2008) 'Welcome to Framenet'. [Online]. Available at: http://framenet.icsi.berkeley.edu/ (Accessed: 20.06.2011).

Frege, G. (1892) 'Über Sinn Und Bedeutung', *Zeitschrift für Philosophie und philosophische Kritik*, 100, pp. 25-50.

Freitag, U. (1992) 'Kartographische Konzepte/Cartographic Conceptions', *Berliner Geowissenchaftliche Abhandlungen*, Reihe C (Band 13),

Garo, L. (2000) 'Introduction to Map Projections'. California State University. [Online]. Available at:

http://personal.uncc.edu/lagaro/cwg/mapproj/intro_mp.html (Accessed: 10.10.2009).

Gaździcki, J. (2002) *Leksykon Geomatyczny – Lexicon of Geomatics.* Warszawa: Polskie Towarzystwo Informacji Przestrzennej.

—. (2005) 'Internetowy Leksykon Geomatyczny'. [Online]. Available at: http://www.ptip.org.pl/auto.php?page=Your_Account (Accessed: 15.08.2011).

Geeraerts, D. (2006) 'Onomasiology and Lexical Variation', in Brown, K. (ed.) *Encyclopedia of Language and Linguistics.* 2nd edn. Amsterdam / Boston: Elsevier, pp. 37-40.

Generalna Dyrekcja Dróg Krajowych i Autostrad (2010) 'Muzeum Techniki Drogowej i Mostowej '. [Online]. Available at: http://www.gddkia.gov.pl/373/muzeum-techniki-drogowej-i-mostowej (Accessed: 16.08.2011).

Ghilani, C. D. & Wolf, P. R. (2008) *Elementary Surveying. An Introduction to Geomatics.* 12th edn. London: Pearson Education.

Gison Sp. z o.o. (2012) 'Dokumentacja Projektu. Świątniki Górne'. [Online]. Available at: http://portal.gison.pl/swiatniki/ (Accessed: 29.10.2012).

Gläser, R. (1998) 'The Stylistic Potential of Phraseological Units in the Light of Genre Analysis', in Cowie, A.P. (ed.) *Phraseology: Theory, Analysis, and Applications.* Oxford: Oxford University Press, pp. 125-143.

Goddard, C. (2002) 'The Search for the Shared Semantic Core of All Languages', in Goddard, C. and Wierzbicka, A. (eds.) *Meaning and Universal Grammar-Theory and Empirical Findings.* Amsterdam / Philadelphia: Benjamins, pp. 5-40.

GOLD (2010) 'General Ontology for Linguistic Description (Gold)'. [Online]. Available at: http://www.linguistics-ontology.org (Accessed: 20.11.2010).

Goraj, S., Garliński, M. & Przybyłowski, K. (1992) *Wytyczne Techniczne G-5.4. Opracowanie Dokumentacji Wyjściowej do Odnowienia Ewidencji Gruntów z Zastosowaniem Technologii Fotogrametrycznej.* Warszawa: Instytut Planowania i Urządzania Terenów Wiejskich Akademii Rolniczo-Technicznej w Olsztynie.

Görög, A. (2006) 'A Database of Cognitive Psychological Terms: Working Method', in ten Hacken, P. (ed.) *Terminology, Computing and Translation.* Tübingen: Narr, pp. 207-220.

Granger, S. & Paquot, M. (2008) 'Disentangling the Phraseological Web', in Granger, S. and Meunier, F. (eds.) *Phraseology. An Interdisciplinary Perspective.* Amsterdam / Philadelphia: Benjamins, pp. 27-50.

Gruber, T. R. (1993) 'A Translation Approach to Portable Ontologies', *Knowledge Acquisition*, 5 (2), pp. 199-220.

—. (2009) 'Ontology', in Liu, L. and Özsu, M.T. (eds.) *Encyclopedia of Database Systems*. New York: Springer-Verlag.

Grzega, J. (2002) 'Some Aspects of Modern Diachronic Onomasiology', *Linguistics: An Interdisciplinary Journal of the Language Sciences*, 40 (5), pp. 1021-1045.

—. (2009) 'Compounding from an Onomasiological Perspective', in Lieber, R. and Štekauer, P. (eds.) *The Oxford Handbook of Compounding*. Oxford: Oxford University Press, pp. 217-232.

Grzega, J. & Schöner, M. (2007) *English and General Historical Lexicology: Materials for Onomasiology Seminars*. Eichstätt: Katholische Universität.

Grzegorczykowa, R., Laskowski, R. & Wróbel, H. (1998) *Gramatyka Współczesnego Języka Polskiego: Morfologia*. 2nd edn. Warszawa: Polskie Wydawnictwo Naukowe.

Guarino, N., Oberle, D. & Staab, S. (2009) 'What Is an Ontology?', in Staab, S. and Studer, R. (eds.) *Handbook on Ontologies*. 2nd edn. Berlin / Heidelberg: Springer.

Hanks, P. (2006) 'Definition', in Brown, K. (ed.) *Encyclopedia of Language and Linguistics*. 2nd edn. Amsterdam / Boston: Elsevier, pp. 399-402.

Hann, M. (2004) *A Basis for Scientific and Engineering Translation: German-English-German* Amsterdam / Philadelphia: Benjamins.

Hartmann, R. R. K. (2006) 'Thesauruses', in Brown, K. (ed.) *Encyclopedia of Language and Linguistics*. 2nd edn. Amsterdam / Boston: Elsevier, pp. 668-676.

Harvey, F. (2005) 'Elasticity for Civil and Political Society between the Formal Cadastre and Informal Land Tenure', *Information Technology for Development*, 12 (4), pp. 291-310.

Heid, U. (2006) 'Extracting Term Candidates from Recursively Chunked Text', in Ten Hacken, P. (ed.) *Terminology, Computing and Translation*. Narr: Tübingen pp. 97-116.

Henssen, J. (1996) 'Basic Principles of the Main Cadastral Systems in the World', *Modern Cadastres and Cadastral Innovations*. Delft.

Hodges, W. (2010) 'Tarski's Truth Definitions'. *Stanford Encyclopedia of Philosophy*. [Online]. Available at: http://plato.stanford.edu/entries/tarski-truth/ (Accessed: 24.11.2010).

Huddleston, R. & Pullum, G., K (2005) *A Student's Introduction to English Grammar*. Cambridge / New York: Cambridge University Press.

Hüllen, W. (1999) 'A Plea for Onomasiology', in Falkner, W. and Schmid, H.-J. (eds.) *Words, Lexemes, Concepts, Approaches to the Lexicon: Studies in Honour of Leonhard Lipka.* Tübingen: Narr, pp. 343-352.

Hycner, R. & Dobrowolska-Wesołowska, M. (2008) *Geodesy, Surveying and Professional Ethics - a Selection of Source Texts with Translation for Students, Lecturers and Surveyors.* Katowice: Gall.

Hycner, R. & Szortyka, I. (2005) *Podręczny Słownik Geodezyjny Angielsko-Polski I Polsko-Angielski (Geodezja i Kartografia, Gospodarka Nieruchomościami i Zagadnienia Prawne, Dydaktyka, Sprawy Studenckie i Funkcjonowanie Uczelni).* Ostrowiec Świętokrzyski: Wydawnictwo Wyższej Szkoły Biznesu i Przedsiębiorczości w Ostrowcu Świętokrzyskim.

Institute of Discrete Mathematics and Geometry of Vienna University of Technology (2008) 'Aspect of Projection'. [Online]. Available at: http://www.geometrie.tuwien.ac.at/karto/aspect.html (Accessed: 14.02.2011).

InvestorWords (2011) 'Investorwords Glossary'. [Online]. Available at: http://www.investorwords.com/1509/dividend.html (Accessed: 17.01.2011).

Irvine, A. D. (2009) 'Rusell's Paradox'. *Stanford Encyclopedia of Philosophy.* [Online]. Available at: http://plato.stanford.edu/entries/russell-paradox/ (Accessed: 24.11.2010).

ISO (1999) *ISO 12620:1999. Computer Applications in Terminology - Data Categories.* Geneva: ISO.

—. (2000a) *ISO 1087-1:2000. Terminology Work - Vocabulary - Part 1: Theory and Application* Geneva: ISO.

—. (2000b) *ISO 9849:2000. Optics and Optical Instruments - Geodetic and Surveying Instruments - Vocabulary.* Geneva: ISO.

—. (2004) *ISO/TR 19122:2004. Geographic Information/Geomatics. Qualification and Certification of Personnel.* Geneva: ISO.

—. (2009) *ISO 704:2009. Terminology Work - Principles and Methods.* Geneva: ISO.

Jackendoff, R. (1983) *Semantics and Cognition.* Cambridge, Mass: MIT Press.

Jackson, H. (1988) *Words and Their Meaning.* Harlow: Longman.

Jagielski, A. (2005) *Geodezja I.* 2nd edn. Kraków: Geodpis.

Janssen, M. (2002) *Simullda: A Multilingual Lexical Database Application Using a Structured Interlingua.* PhD thesis. Universiteit Utrecht.

—. (2004) 'Multilingual Lexical Detabases, Lexical Gaps, and Simullda', *International Journal of Lexicography*, 17 (2), pp. 137-154.

Jespersen, O. (1942) *A Modern English Grammar. Part VI, Morphology.* Copenhagen: Ejnar Munksgaard.

Johnson, J. M. (2003) *Geographic Information.* Westport, CT: Greenwood.

Kameyama, M., Ochitani, R. & Peters, S. (1991) 'Resolving translation Mismatches with Information Flow', *Proceedings of the 29th Meeting of the Association of Computational Linguistics.* Berkley, pp. 193-200.

Kameyama, M., Peters, S. & Schütze, H. (1993) 'Combining Logic-Based and Corpus-Based Methods for Resolving Translation Mismatches', *Working Notes of the Aaai Spring Symposium on Building Lexicons for Machine Translation.* Menlo Park CA: AAAI Press pp. 92-98.

Kancelaria Sejmu RP (1989) Ustawa z dnia 17 Maja 1989 r. Prawo Geodezyjne i Kartograficzne. Warszawa.

Karpova, O. (2006) 'Russian Lexicography', in Brown, K. (ed.) *Encyclopedia of Language and Linguistics.* 2nd edn. Amsterdam / Boston: Elsevier, pp. 704-715.

KARTO (2005) 'Słownik Pojęć Kartograficznych'. [Online]. Available at: http://www.karto.pl/slownik/kartografia (Accessed: 02.10.2009).

Kasprzak, M. (2009) 'Metody Prezentacji Danych'. gisplay.pl. [Online]. Available at: http://www.gisplay.pl/kartografia/metody-prezentacji-danych.html (Accessed: 10.10.2009).

Katamba, F. (1993) *Morphology.* Basingstoke: Macmillan.

Kent County Council (2011) 'About Public Rights of Way'. [Online]. Available at: http://kent.gov.uk/environment_and_planning/country side_access/about_public_rights_of_way/using_public_rights_of_way. aspx (Accessed: 18.07.2011).

Klemensiewicz, Z. (1962) *Podstawowe Wiadomości z Gramatyki Języka Polskiego.* 2nd edn. Warszawa: Polskie Wydawnictwo Naukowe.

Knowles, M. & Moon, R. (2006) *Introducing Metaphor.* London/New York: Routledge.

Kozłowski, J. (1997) 'Polish Experience in Creation of Real Property Registration System', *High-level Technical Seminar: Private and Public Sector Cooperation in National Land Tenure Development in Eastern and Central Europe*. Bertinoro, Italy: 1-5 April 1997 Food and Agriculture Organization of the United Nations. Available at: http://www.fao.org/sd/LTdirect/LTforum/LTfo0011.htm (Accessed: 1.11.2011).

Kraak, M.-J. & Ormeling, F. (2003) *Cartography. Visualisation of Geospatial Data.* 2nd edn. Harlow: Pearson Education.

Kreidler, C. W. (1998) *Introducing English Semantics.* London / New York: Routledge.
Kunach, W. (1999) 'Rejestr Nieruchomości'. *Praktyka.* [Online]. Available at: http://www.nieruchomosci.beck.pl/index.php?mod=m_artykuly &cid=16&id=317 (Accessed: 26.01.2010).
Kurcz, I., Lewicki, A., Szfran, K. & Woronczak, J. (1990) *Słownik Frekwencyjny Polszczyzny Współczesnej- SFPW [a Frequency Dictionary of Contemporary Polish].* Kraków: Instytut Języka Polskiego PAN.
Kwiatek, E. & ten Hacken, P. (2011) 'Efficiency of Multiterm for Extraction of Terms from English and Polish Specialized Corpora', in Goźdź-Roszkowski, S. (ed.) *Explorations across Languages and Corpora.* Łódź Studies in Language. Frankfurt am Main: Peter Lang, pp. 145-162.
Labov, W. (1973) 'The Boundaries of Words and Their Meanings', in Bailey, C.J.N. and Shuy, R.W. (eds.) *New Ways of Analyzing Variation in English.* Washington DC: Georgetown University Press, pp. 340-373.
Lakoff, G. (1987) *Women, Fire and Dangerous Things: What Categories Reveal About the Mind.* Chicago: University of Chicago Press.
Lakoff, G. & Johnson, M. (1980) *Metaphors We Live By.* Chicago / London: University of Chicago Press.
Lakoff, G. & Johnson, M. (1999) *Philosophy in the Flesh: The Embodied Mind and Its Challenge to Western Thought.* New York: Basic Books.
Lamy, M.-N. & Klarskov Mortensen, H. J. (2011) 'Using Concordance Programs in the Modern Foreign Languages Classroom'. *ICT4LT (Information and Communications Technologies for Language Teachers) Module.* [Online]. Available at: http://www.ict4lt.org/en/en_mod2-4.htm (Accessed: 14.07.2011).
Land Register Online (2008) 'Title Deeds'. [Online]. Available at: http://www.landregistrydeeds.co.uk/Title%20Deeds/Title_Deeds.htm (Accessed: 15.08.2011).
Land Registry (2008a) 'Example of the Title Plan. Title Number: CS72510.' [Online]. Available at: http://www.landregistry.gov.uk/www/wps/QDMPS-Portlet/resources/example_title_plan.pdf (Accessed: 15.08.2011).
—. (2008b) 'Extracts from Title Register. Title Number: CS72510.' [Online]. Available at: http://www.landregistry.gov.uk/www/wps/QDMPS-portlet/resources/example_register.pdf (Accessed: 15.08.2011).
—. (2008c) 'Glossary'. [Online]. Available at: http://www.landregistry.gov.uk/ (Accessed: 15.08.2011).

—. (2012a) '6 Appendix a - Example of a Register'. Land Registry. [Online]. Available at: http://www.landregistry.gov.uk/public/guides/public-guide-1#guide-mark-22 (Accessed: 15.08.2011).
—. (2012b) '7 Appendix B - Example of a Title Plan'. Land Registry. [Online]. Available at: http://www.landregistry.gov.uk/public/guides/public-guide-1#guide-mark-22 (Accessed: 15.08.2011).
Langacker, R. W. (1987) *Foundations of Cognitive Grammar.* Stanford, California: Stanford University Press.
Lawrence, V. (2011) 'Ordnance Survey Glossary'. Ordnance Survey. [Online]. Available at: http://www.ordnancesurvey.co.uk/oswebsite/aboutus/reports/misc/glossary.html (Accessed: 24.02.2011).
Leech, G. (1991) 'The State of the Art in Corpus Linguistics', in Aijmer, K. and Atltenberg, B. (eds.) *English Corpus Linguistics.* London/New York: Longman, pp. 8-29.
Lees, R. (1960) *The Grammar of English Nominalizations.* reissued 1963, 5th printing 1968 edn. Den Haag: Indiana University, Bloomington & Mouton.
Lieber, R. (2005) 'English Word Formation Processes', in Štekauer, P. and Lieber, R. (eds.) *Handbook of Word-Formation* Dordrecht: Springer, pp. 375-428.
—. (2009a) 'Ie, Germanic: English', in Lieber, R. and Štekauer, P. (eds.) *The Oxford Handbook of Compounding.* Oxford: Oxford University Press, pp. 357-369.
—. (2009b) 'A Lexical Semantic Approach to Compounding', in Lieber, R. and Štekauer, P. (eds.) *The Oxford Handbook of Compounding.* Oxford: Oxford University Press, pp. 78-104.
Longworth, G. (2006) 'Definitions: Uses and Varieties Of', in Brown, K. (ed.) *Encyclopedia of Language and Linguistics.* 2nd edn. Amsterdam/Boston: Elsevier, pp. 409-412.
Lowe, J. B., Baker, C. F. & Fillmore, C. J. (1997) 'A Frame-Semantic Approach to Semantic Annotation', *Siglex Workshop on Tagging Text with Lexical Semantics: Why, What, and How?, Washington, D.C., in Conjunction with the Fifth Conference on Applied Natural Language Processing (Anlp 97).* Washington, D.C.
Lüdeling, A., Schmid, T. & Kiokpasoglou, S. (2002) 'Neoclassical Word Formation in German', in Booij, G. and van Marle, J. (eds.) *Yearbook of Morphology 2001.* Dordrecht: Kluwer, pp. 253-283.
Lyman, J. & Wright, J. W. (2009) 'Surveying', *Encyclopaedia Britannica.* [CD-ROM].

Lyons, J. (1977) *Semantics.* vol. 1. Cambridge: Cambridge University Press

Madsen, B. N. (2007) 'Ontologies and Indeterminacy', in Antia, B. (ed.) *Indeterminacy in Terminology and LSP Studies in Honour of Heribert Picht.* Amsterdam / Philadelphia: Benjamins, pp. 180-198.

Madsen, B. N. & Thomsen, H. E. (2009) 'Caos - a Tool for the Construction of Terminological Ontologies', in Jokinen, K. and Bick, E. (eds.) *Nordic Conference of Computational Linguistics (Nodalida)* Odense, Denmark, pp. 279-282.

Madsen, B. N., Thomsen, H. E. & Vikner, C. (2006) 'Repræsentation Af Inddelingskriterier I Caos 2', in Ágústa, P. (ed.) *Nordterm 14.* Reykjavik Þorbergsdóttir: Ord og Termer, pp. 274-287.

Maine Society of Land Surveyors (n.d) 'Boundary Surveys'. [Online]. Available at: http://www.msls.org/index.html (Accessed: 20.06.2010).

Mani, I. & Maybury, M. T. (1999) *Advances in Automatic Text Summarization.* Cambridge, Mass.: MIT Press.

Manning, C. D. & Schütze, H. (1999) *Foundations of Statistical Natural Language Processing.* Cambridge, Mass: MIT Press.

Marchand, H. (1969) *The Categories and Types of Present-Day English Word-Formation: A Synchronic-Diachronic Approach.* 2nd edn. München: Beck.

McEnery, T. & Wilson, A. (2006) *Corpus Linguistics.* 2nd edn. Edinburgh: Edinburgh University Press.

Mizerski, W., Dulna-Rak, E., Rucińska, A., Kipert, K., Kiepas, A. & Bednarczuk, B. (2000) *Język Polski. Encyklopedia w Tabelach.* Warszawa: Adamantan.

Montague, R. (1970) 'Universal Grammar', *Theoria*, 36 pp. 373–398.

National Parks and Access to the Countryside Act 1949.

National Trails (2011) 'British National Trails'. [Online]. Available at: http://www.nationaltrail.co.uk/text.asp?PageId=2 (Accessed: 15.08.2011).

Natural Resources Canada (2012) 'Map Projections'. [Online]. Available at: http://atlas.nrcan.gc.ca/site/english/learningresources/carto_corner /map_projections.html/document_view (Accessed: 20.10.2010).

Naturenet (2011) 'Naturenet - the Uk Countryside and Nature Conservation'. [Online]. Available at: http://www.naturenet.net/row /reclass.html (Accessed: 15.08.2011).

Newmark, P. (1988) *A Textbook of Translation.* New York: Prentice Hall.

NGS (2010) 'National Geodetic Survey'. [Online]. Available at: http://www.ngs.noaa.gov/ (Accessed: 13.10.2010).

Nuessel, F. (2006) 'Semiology Vs. Semiotics', in Brown, K. (ed.) *Encyclopedia of Language and Linguistics.* 2nd edn. Amsterdam / Boston: Elsevier, pp. 193-194.

Ogden, C. K. & Richards, I. A. (1949) *The Meaning of Meaning: A Study of the Influence of Language Upon Thought and of the Science of Symbolism* 10th edn. New York: Mariner Books.

Ogorzelska, B. (2006) 'Odwzorowania Kartograficzne', in Pasławski, J. (ed.) *Wprowadzenie do Kartografii i Topografii.* Wrocław: Nowa Era, pp. 81-168.

Okła, K. (2009) 'Encyklopedia Leśna- Dział Geodezja i Geomatyka'. [Online]. Available at: http://www.encyklopedialesna.pl/hasla.php (Accessed: 10.10.2010).

OSP Project Software (2010) [Cd-Rom]. Surveying.

Ośrodek Usług Inżynierskich STAAND Sp. z o.o. (2012) *Dokumentacja Projektu. Elektrownia Wodna Sieniawa,* Kraków: Staand.

Pasławski, J. (2006a) 'Kartograficzne Metody Prezentacji', in Pasławski, J. (ed.) *Wprowadzenie do Kartografii i Topografii.* Wrocław: Nowa Era, pp. 196-234.

Pasławski, J. (ed.) (2006b) *Wprowadzenie Do Kartografii I Topografii.* Wrocław: Nowa Era.

Pearson, J. (1998) *Terms in Context.* Amsterdam / Philadelphia: Benjamins.

Permanent Committee on Cadastre in the European Union (2011) 'Cadastre in the United Kingdom'. [Online]. Available at: http://www.eurocadastre.org/eng/coordinationclausuk.html (Accessed: 15.08.2011).

Petropoulou, E. & ten Hacken, P. (2002) 'Neo-Classical Word Formation in WM Electronic Dictionaries', *Proceedings of the Tenth Euralex International Congress.* Copenhagen, pp. 169-174.

Piasecki, M., Szpakowicz, S. & Broda, B. (2009) *A Wordnet from the Ground Up.* Wrocław: Oficyna Wydawnicza Politechniki Wrocławskiej.

Pietroski, P. (2009) 'Logical Form'. *Stanford Encyclopedia of Philosophy.* [Online]. Available at: http://plato.stanford.edu/entries/logical-form/ (Accessed: 24.11.2010).

Plag, I. (2003) *Word-Formation in English.* Cambridge: Cambridge University Press.

Plag, I., Braun, M., Lappe, S. & Schramm, M. (2007) *Introduction to English Linguistics* Berlin: Mouton de Gruyter.

Podhajecka, M. (2002) 'Zapożyczenia Polskie w Języku Angielskim na Materiale Oxford English Dictionary (OED)', *Język Polski,* 82 (5), pp. 330-337.

Polish WordNet (2010) 'Semi-Automatic Construction of Plwordnet'. Wroclaw University of Technology. [Online]. Available at: http://www.plwordnet.pwr.wroc.pl/main/?lang=en (Accessed: 9.08.2011).

Polska Organizacja Turystyczna (2011) 'Turystyczne Znaki Drogowe'. [Online]. Available at: http://www.pot.gov.pl/drogowe-znaki-turystyczne/turystyczne-znaki-drogowe/ (Accessed: 20.07.2011).

Polski Komitet Normalizacyjny (1986) *PN-N-02207:1986 Geodezja. Terminologia.* Warszawa: Ars Boni Sp. z o.o.

Prahl, B. & Pretzolt, S. (1997) 'Translation Problems and Translation Strategies Involved in Human and Machine Translation: Empirical Studies', in Hauenschild, C. and Heizmann, S. (eds.) *Machine Translation and Translation Theory.* Berlin / New York: Mouton de Gruyter, pp. 123-144.

Princeton University (2010) 'About Wordnet'. Princeton University. [Online]. Available at: http://wordnet.princeton.edu/ (Accessed: 24.01.2011).

Procter, P. (ed.) (1978) *Longman Dictionary of Contemporary English.* Harlow: Longman.

Protégé (2010) 'Protégé '. Stanford Centre for Biomedical Information Research. [Online]. Available at: http://protege.stanford.edu/ (Accessed: 5.01.2011).

Przepiórkowski, A. (2008) 'Korpus Języka Polskiego Instytutu Podstaw Informatyki Polskiej Akademii Nauk [IPI PAN Corpus - Polish Corpus Developed at the Institute of Computer Science, Polish Academy of Sciences – ICS PAS]'. [Online]. Available at: http://korpus.pl/index.php?page=welcome (Accessed: 26.07.2009).

PWN (2009a) 'PWN Corpus of Polish'. [Online]. Available at: http://korpus.pwn.pl/index_en.php (Accessed: 26.07.2009).

—. (2009b) 'Word Frequency in the Demonstration Corpus'. [Online]. Available at: http://korpus.pwn.pl/stslow_en.php (Accessed: 23.03.2009).

—. (2010) 'Encyklopedia PWN'. Available at: http://encyklopedia.pwn.pl/ (Accessed: 11.08.2010).

Quine, W. V. O. (1961) *Word and Object.* Cambridge, Mass.: MIT Press.

Ramblers (2011) 'Access for Walkers in Britain'. [Online]. Available at: http://www.ramblers.org.uk/info/britain/access-for-walkers-in-britain.htm (Accessed: 10.08.2011).

Ratajski, L. (1989) *Metodyka Kartografii Społeczno - Gospodarczej.* Warszawa / Wrocław: Państwowe Przedsiębiorstwo Wydawnictw Kartograficznych.

Reimer, M. (2009) 'Reference'. *Stanford Encyclopedia of Philosophy*. [Online]. Available at: http://plato.stanford.edu/entries/reference/#Rel (Accessed: 11.10.2009).

Riemer, N. (2010) *Introducing Semantics*. Cambridge: Cambridge University Press.

Ritter, M. (2006) 'The Physical Environment: An Introduction to Physical Geography'. [Online]. Available at: http://www.uwsp.edu/geo/faculty/ritter/geog101/textbook/title_page.html (Accessed: 16.03.2011).

RMRRB (2001) *Rozporządzenie Ministra Rozwoju Regionalnego i Budownictwa z dnia 29 Marca 2001 r. w sprawie Ewidencji Gruntów i Budynków*.

Road Traffic Act 1988.

Rosch, E. (1975) 'Cognitive Representation of Semantic Categories', *Journal of Experimental Psychology*, 104, pp. 192-233.

Roy, S. K. (2010) *Fundamentals of Surveying*. 2nd edn. New Delhi: Prentice Hall.

Runge, A. & Runge, J. (2009) *Słownik Pojęć z Geografii Społeczno-Ekonomicznej*. Chorzów: Videograf Edukacja.

Russel, B. (1905) 'On Denoting', *Mind*, 14, pp. 479-493.

Saeed, J. I. (2003) *Semantics*. 2nd edn. Oxford: Blackwell Publishing.

Sag, I. A., Baldwin, T., Bond, F., Copestake, A. A. & Flickinger, D. (2002) 'Multi-Word Expressions: A Pain in the Neck for NLP', *Proceedings of the Third International Conference on Intelligent Text Processing and Computational Linguistics (CICLING 2002)*. Mexico City, pp. 1-15.

Sager, J. C. (1990) *A Practical Course in Terminology Processing*. Amsterdam / Philadelphia: Benjamins.

—. (2001) 'Terminology Compilation: Consequences and Aspects of Automation', in Wright, S.E. and Budin, G. (eds.) *Handbook of Terminology Management*. Amsterdam / Philadelphia: Benjamins, pp. 761-771.

Saussure, F. (1916/1969) *Cours de Linguistique Générale*. Translated by Baskin, W., Paris: Payot.

Scalise, S. & Bisetto, A. (2009) 'The Classification of Compounds', in Lieber, R. and Štekauer, P. (eds.) *The Oxford Handbook of Compounding*. Oxford: Oxford University Press, pp. 34-54.

Schmidt, T. (n.d) 'Kicktionary. The Multilingual Electronic Dictionary of Football Language'. [Online]. Available at: http://www.kicktionary.de/index.html (Accessed: 24.06.2011).

Schmitter, P. (2008) 'The Theory of Word Formation in Early Semasiology: A Blank Spot on the Map of 19th Century Linguistics', *Language Sciences*, 30 (5), pp. 575-596.

Schmitz, K.-D. (2001) 'Criteria for Evaluating Terminology Database Management Programs', in Wright, S.E. and Budin, G. (eds.) *Handbook of Terminology Management.* Amsterdam / Philadelphia: Benjamins, pp. 539-551.

—. (2006) 'Terminology and Terminological Databases', in Brown, K. (ed.) *Encyclopedia of Language and Linguistics.* 2nd edn. Amsterdam/Boston: Elsevier, pp. 578-587.

—. (2008a) 'Applied Principles of Terminology Work'. *International Terminology Summet School 2008.* [presentation] Vienna: Univeristy of Vienna, Center for Translation Sciences.

—. (2008b) 'A Closer Look at Terminology Management Systems'. *International Terminology Summer School 2008.* [presentation] Vienna: Univeristy of Vienna, Center for Translation Sciences.

—. (2008c) 'Data Modelling for Terminology Management'. *International Terminology Summer School 2008.* [presentation] Vienna: University of Vienna, Center for Translation Sciences.

Scott, M. (2010a) 'Wordsmith Tools'. Lexical Analysis Software Ltd. [Online]. Available at: http://www.lexically.net/wordsmith/index.html (Accessed: 14.07.2011).

—. (2010b) *Wordsmith Tools Step by Step.* Liverpool: Lexical Analysis Software Ltd.

SDL MultiTerm Extract: Tools Guide 2007 [Computer Program].

SDL Trados (2007) [Computer Program].

Sękowska, E. (2002) 'Budowa Wyrazów', in Dubisz, S. (ed.) *Nauka o Języku dla Polonistów.* 4th edn. Warszawa: Książka i Wiedza, pp. 156-187.

Simpson, S. R. (1976) *Land Law and Registration.* Cambridge: Cambridge University Press.

Sinclair, J. (1995) *Corpus, Concordance, Collocation.* Oxford, New York: Oxford University Press.

Skinner, B. F. (1957) *Verbal Behaviour.* Ann Arbor, MI: Copley Publishing Group.

Smith, R. J. (2010) *Introduction to Land Law.* 2nd edn. New York: Pearson Longman.

Snowdon, P. (2009) 'Peter Frederick Strawson'. *Stanford Encyclopedia of Philosophy*. [Online]. Available at: http://plato.stanford.edu/entries/strawson/ (Accessed: 24.11.2010).

Štekauer, P. (2006) 'Onomasiological Theory of Word-Formation', in Brown, K. (ed.) *Encyclopedia of Language and Linguistics.* 2nd edn. Amsterdam/Boston: Elsevier, pp. 34-37.

Stern, G. (1931) *Meaning and Change of Meaning with Special Reference to the English Language.* Göteborg: Elander

Stevens, R., Goble, C. A. & Bechhofer, S. (2000) *Ontology-Based Knowledge Representation for Bioinformatics.* Manchester: Department of Computer Science and School of Biological Science, University of Manchester Press.

Strawson, P. F. (1950) *On Referring.* London: Blackwell.

Sydenham, A. (2001) *Public Rights of Way and Access to Land.* Bristol: Jordan.

Szabó, Z. G. (2007) 'Compositionality'. *Stanford Encyclopedia of Philosophy.* [Online]. Available at: http://plato.stanford.edu/entries/compositionality/#1.5.4 (Accessed: 24.11.2010).

Szejba, A., Grzechnik, B., Kacprzak, S., Kulka, J. & Musiatowicz, H. (1983) *Instrukcja Techniczna G-4, Pomiary Sytuacyjne i Wysokościowe.* Warszawa: GUGiK.

Szymanek, B. (1988) *Categories and Categorization in Morphology.* Lublin: Redakcja Wydawnictw KUL.

—. (2009) 'Ie, Slavonic: Polish', in Lieber, R. and Štekauer, P. (eds.) *The Oxford Handbook of Compounding.* Oxford: Oxford University Press, pp. 464-477.

Talmy, L. (1981) 'Force Dynamics', *Paper presented at conference on Language and Mental Imagery.* University of California, Berkeley.

—. (1985) 'Lexicalization Patterns: Semantic Structure in Lexical Forms. Language Typology and Syntactic Description', in Shopen, T. (ed.) *Grammatical Categories and the Lexicon.* Cambridge: Cambridge University Press.

—. (1988) 'The Relation of Grammar to Cognition', in Rudzka-Ostyn, B. (ed.) *Topics in Cognitive Linguistics.* Amsterdam / Philadelphia: Benjamins, pp. 165-205.

—. (2000) *Toward a Cognitive Semantics.* Cambridge, Mass: MIT Press.

Tarski, A. (1933) *Pojęcie Prawdy w Językach Nauk Dedukcyjnych.* Warszawa: Towarzystwo Naukowe Warszawskie.

Tatarczyk, J. (1991) *Słownik Geodezyjny Polsko-Angielsko-Niemiecki.* Kraków: Wydawnictwo AGH.

—. (2005) *Słownik Geodezyjny Polsko-Angielsko-Niemiecki. [CD-ROM].* Katowice: Gall.

Temmerman, R. (2000) *Towards New Ways of Terminology Description: The Sociocognitive Approach* Amsterdam / Philadelphia: Benjamins.

ten Hacken, P. (1994) *Defining Morphology : A Principled Approach to Determining the Boundaries of Compounding, Derivation, and Inflection.* Informatik und Sprache. Hildesheim: G. Olms Verlag.

—. (2010a) 'Creating Legal Terms: A Linguistic Perspective', *International Journal for the Semiotics of Law*, 23 (4), pp. 407-425.

—. (2010b) 'The Tension between Definition and Reality in Terminology', in Dykstra, A. and Schoonheim, T. (eds.) *Proceedings of the XIV Euralex International Congress.* Leeuwarden: Fryske Akademy.

Teubert, W. (1996) 'Comporable or Parallel Corpora?', *International Journal of Lexicography*, 9 (3), pp. 238-264.

The Atlas of Canada (2009) 'Glossary'. [Online]. Available at: http://atlas.nrcan.gc.ca/site/english/learningresources/glossary/results.html?term= (Accessed: 24.02.2011).

Thelen, M. & Steurs, F. (2010) *Terminology in Everyday Life.* Amsterdam / Philadelphia: Benjamins.

Tognini-Bonelli, E. (1996) *Corpus Theory and Practice.* Birmingham: TWC.

Trask, R. L. (1999) *Key Concepts in Language and Linguistics.* London/New York: Routledge.

Tyner, J. A. (2010) *Principles of Map Design.* New York: The Guilford Press.

UCAS (2011) 'Universities & Colleges Admissions Service (Ucas)'. [Online]. Available at: http://www.ucas.com/ (Accessed: 25.06.2011).

UDC Consortium (2010) 'Universal Decimal Classification (Udc)'. UDC Consortium. [Online]. Available at: http://udcc.org/ (Accessed: 25.06.2011).

Ullman, S. (1962) *Semantics: An Introduction to the Science of Meaning* 2nd edn. Oxford: Basil Blackwell.

Uren, J. & Price, W. F. (2010) *Surveying for Engineers.* 5th edn. Basingstoke: Palgrave Macmillan.

Ustawa z dnia 16 kwietnia 2004 o Ochronie Przyrody Warszawa: Kancelaria Sejmu.

van der Vliet, H. (2006) 'Combinatorics for Special Purposes', in ten Hacken, P. (ed.) *Terminology, Computing and Translation.* Tübingen: Narr, pp. 57-72.

—. (2007) 'The Referentiebestand Nederlands as a Multi-Purpose Lexical Database', *International Journal of Lexicography*, 20 (3), pp. 239-257.

Vinay, J.-P. & Darbelnet, J. (1995) *Comparative Stylistics of French and English: A Methodology for Translation.* Translated by Sager, J.C. and Hamel, M.-J., Amsterdam / Philadelphia: Benjamins.

Vossen, P. (1997) 'EuroWordNet: A Multilingual Database for Information Retrieval', in Sheridan, P. (ed.) *Proceedings of the DELOS Workshop on Cross-Language Information Retrieval.* Zurich: ATM.

Vossen, P. (ed.) (1998) *EuroWordNet: A Multilingual Database with Lexical Semantic Networks*. Dordrecht: Kluwer.

Vossen, P. (2009) 'From WordNet, to EuroWordNet, to the Global Wordnet Grid: Anchoring Languages to Universal Meaning', *Guest Lecture, Language Engineering Applications, February, 26th 2009, Leuven.*

Warburton, K. (2001) 'Globalization and Terminology Management ', in Wright, S.E. and Budin, G. (eds.) *Handbook of Terminology Management.* Amsterdam / Philadelphia: Benjamins, pp. 677-696.

Whalen, D. (2004) 'How the Study of Endangered Languages Will Revolutionize Linguistics ', in van Sterkenburg, P. (ed.) *Linguistics Today – Facing a Greater Challenge.* Amsterdam / Philadelphia: Benjamins, pp. 321-342.

Wierzbicka, A. (1972) *Semantic Primitives.* Frankfurt: Athenäum.

—. (1984) '"Apples" Are Not A "Kind of Fruit": The Semantics of Human Categorization', *American Ethnologist*, 11 (2), pp. 313-328.

—. (1990) 'The Meaning of Color Terms: Semantics, Culture, and Cognition', *Cognitive Linguistics 1*, pp. 99-150.

—. (1996) *The Primitives of Linguistic Meaning. Semantics: Primes and Universals.* Oxford: Oxford University Press.

Wiese, R. (1996) 'Phrasal Compounds and the Theory of Word Syntax', *Linguistic Inquiry*, 27, pp. 183-193.

Wildlife and Countryside Act 1981.

Winston, M. E., Chaffin, R. & Hermann, D. J. (1987) 'A Taxonomy of Part-Whole Relations', *Cognitive Science*, 11 (4), pp. 417-444.

Witalisz, A. (2006) 'English Linguistic Influence on Polish and Other Slavonic Languages', *Quaderni Del Ceslic. Occasional Papers.* Bologna: Centro di Studi Linguistico-Culturali (CeSLiC).

—. (2007) 'Some Remarks on Hidden Anglicisms'. [Paper presented at 38th Poznań Linguistic Meeting] Gniezno.

Wittgenstein, L. (1953) *Philosophical Investigations* Translated by Anscombe, G.E.M., London: Blackwell.

Wołłodko, Z. (1973) *Geodezja.* Warszawa: Państwowe Wydawnictwo Rolnicze i Leśne.

Worldmapper (2006) 'Total Population Cartogram'. [Online]. Available at: http://www.worldmapper.org/index.html (Accessed: 10.10.2010).

Wright, J. (1982) *Ground and Air Survey for Field Scientists.* Oxford: Clarendon.

Wright, S. E. (1997) 'Term Selection: The Initial Phase of Terminology Management', in Wright, S.E. and Budin, G. (eds.) *Handbook of Terminology Management.* Amsterdam / Philadelphia: Benjamins, pp. 13-23.

—. (2001a) 'Data Catagories for Terminology Management ', in Wright, S.E. and Budin, G. (eds.) *Handbook of Terminology Management.* Amsterdam / Philadelphia: Benjamins, pp. 552-571.

—. (2001b) 'Terminology Management Entry Structures ', in Wright, S.E. and Budin, G. (eds.) *Handbook of Terminology Management.* Amsterdam / Philadelphia: Benjamins, pp. 572-599.

Wright, S. E. & Budin, G. (eds.) (2001) *Handbook of Terminology Management.* Volume 2: Application-Oriented Terminology Management Amsterdam / Philadelphia: Benjamins.

Wroclaw University of Technology (2006) 'Semi-Automatic Construction of Plwordnet'. [Online]. Available at: http://www.plwordnet.pwr.wroc.pl/main/?lang=en (Accessed: 9.08.2011).

Wüster, E. (1931) *Internationale Sprachnormung in der Technik, Besonders in der Elektrotechnik.* PhD Thesis. VDJ.

—. (1969) 'Die Vier Dimensionen der Terminologiearbeit', *Mitteilungsblatt fur Dolmetscher und Übersetzer*, 15 (2), pp. 1-6.

—. (1979) *Einführung in Die Allgemeine Terminologielehre Und Terminologische Lexikographie.* Vienna/New York: Springer.

Wyler, S. (1992) *Colour and Language: Colour Terms in English* Tübingen: Narr.

Yule, G. (2010) *The Study of Language.* 4th edn. Cambridge: Cambridge University Press.

Zabawa, M. (2008) 'English-Polish Language Contact and Its Influence on the Semantics of Polish', in Kątny, A. (ed.) *Kontakty Językowe i Kulturowe w Europie / Sprach- und Kulturkontakte in Europa.* Gdańsk: Wydawnictwo Uniwersytetu Gdańskiego, pp. 154-164.

Zawisławska, M. (ed.) (2010) *Ramki: Rygorystyczna Aplikacja Metodologii Kognitywno Interpretacyjnej* Warszawa: Elipsa.

INDEX

U

V

W